INFORMATION SYSTEMS AND THE ENVIRONMENT

Edited by

Deanna J. Richards

Braden R. Allenby

and

W. Dale Compton

NATIONAL ACADEMY OF ENGINEERING

NATIONAL ACADEMY PRESS
Washington, D.C.

NATIONAL ACADEMY PRESS • 2101 Constitution Avenue, N.W. • Washington, D.C. 20418

The National Academy of Engineering was established in 1964, under the charter of the National Academy of Sciences, as a parallel organization of outstanding engineers. It is autonomous in its administration and in the selection of its members, sharing with the National Academy of Sciences the responsibility for advising the federal government. The National Academy of Engineering also sponsors engineering programs aimed at meeting national needs, encourages education and research, and recognizes the superior achievements of engineers. Dr. Wm. A. Wulf is president of the National Academy of Engineering.

This volume has been reviewed according to procedures approved by a National Academy of Engineering report review process. The interpretations and conclusions expressed in the papers are those of the authors and are not presented as the views of the council, officers, or staff of the National Academy of Engineering.

This activity was undertaken in partnership with the H. John Heinz III Center for Science, Economics and the Environment.

International Standard Book Number: 0-309-06243-8

Library of Congress Catalog Card Number 2001093506

Copies of this report are available from National Academy Press, 2101 Constitution Avenue, N.W., Lockbox 285, Washington, D.C. 20055; (800) 624-6242 or (202) 334-3313 (in the Washington metropolitan area); Internet, http://www.nap.edu

Cover Art: *Untitled* (detail) by Maurice Golubov, courtesy of Smithsonian American Art Museum, Gift of Patricia and Phillip Frost.

THE NATIONAL ACADEMIES

National Academy of Sciences
National Academy of Engineering
Institute of Medicine
National Research Council

The **National Academy of Sciences** is a private, nonprofit, self-perpetuating society of distinguished scholars engaged in scientific and engineering research, dedicated to the furtherance of science and technology and to their use for the general welfare. Upon the authority of the charter granted to it by the Congress in 1863, the Academy has a mandate that requires it to advise the federal government on scientific and technical matters. Dr. Bruce M. Alberts is president of the National Academy of Sciences.

The **National Academy of Engineering** was established in 1964, under the charter of the National Academy of Sciences, as a parallel organization of outstanding engineers. It is autonomous in its administration and in the selection of its members, sharing with the National Academy of Sciences the responsibility for advising the federal government. The National Academy of Engineering also sponsors engineering programs aimed at meeting national needs, encourages education and research, and recognizes the superior achievements of engineers. Dr. William A. Wulf is president of the National Academy of Engineering.

The **Institute of Medicine** was established in 1970 by the National Academy of Sciences to secure the services of eminent members of appropriate professions in the examination of policy matters pertaining to the health of the public. The Institute acts under the responsibility given to the National Academy of Sciences by its congressional charter to be an adviser to the federal government and, upon its own initiative, to identify issues of medical care, research, and education. Dr. Kenneth I. Shine is president of the Institute of Medicine.

The **National Research Council** was organized by the National Academy of Sciences in 1916 to associate the broad community of science and technology with the Academy's purposes of furthering knowledge and advising the federal government. Functioning in accordance with general policies determined by the Academy, the Council has become the principal operating agency of both the National Academy of Sciences and the National Academy of Engineering in providing services to the government, the public, and the scientific and engineering communities. The Council is administered jointly by both Academies and the Institute of Medicine. Dr. Bruce M. Alberts and Dr. William A. Wulf are chairman and vice chairman, respectively, of the National Research Council.

Preface

Today, the knowledge base on which environmental decisions are being made is broader and deeper than ever before. Information technology has introduced new opportunities for harnessing such knowledge to improve environmental performance. That, in part, is the subject of this volume of papers. The book also speculates about the potential contribution of information technology to sustainable development.

Few will argue that increased knowledge will play an essential role in meeting humanity's environmental challenges. Yet, much of the quest for the knowledge that is needed falls into the category of several "public goods" challenges that no single company can justify undertaking alone and which can have a dramatic payoff if companies can share costs and responsibilities or if the government were to step up to the plate and fill the void. The challenges range from articulating technical and management standards that reflect best strategic environmental approaches and defining criteria for determining environmental impacts and metrics of environmental performance, to the potential use and misuse of environmental information. In each of these areas, there are important roles for government, trade associations, industry, universities and environmental public interest groups (preferably working collaboratively).

This volume originated from a July 1997 workshop conducted in partnership with the H. John Heinz III Center for Science, Economics and the Environment. Both the publication and the meeting are components of the NAE's program on Technology and Sustainable Development. We are indebted to the authors for their excellent contributions, to Robert M. White for his review of those contributions, and to an editorial team composed of Penny Gibbs, Greg Pearson, Long Nguyen, Deanna J. Richards, and Karla J. Weeks. Special thanks also go to Brad Allenby and Dale Compton for their efforts in chairing the workshop and for their contributions to the overview and perspectives in this volume.

Wm. A. Wulf
President

Contents

OPPORTUNITIES FOR COLLABORATION
AND NEW TECHNOLOGIES

INFORMATION SYSTEMS AND THE ENVIRONMENT

Information Systems and the Environment
Overview and Perspectives

BRADEN R. ALLENBY, W. DALE COMPTON,
and DEANNA J. RICHARDS

Today, solutions to environmental challenges are aided by an arsenal of information and knowledge systems that were unavailable for most of the last 30 years when environmental management was predicated on "command and control" mechanisms such as remediation of specific sites or compliance with, and enforcement of, end-of-pipe emissions requirements and standards. As knowledge about the causes of environmental ills has grown, so too has the number of options on how to handle them and the development of collaborations and partnerships aimed at harnessing the growing incentive-based approaches to environmental protection. As additional information technologies and knowledge management techniques evolve, environmental considerations will join other areas of strategic importance to industry.

Information technologies are unique not just because of their growing use in decision-making and knowledge management systems, important as that is. Their use has also yielded significant improvements in the efficiency of energy and materials use. This has contributed to economic expansion without the increases in environmental impacts that would have resulted had the efficiency improvements not occurred. Advances in information technology are likely to continue to provide opportunities for the development of improved and new products and services.

This will not occur, however, without continuing attention to both the individual units (e.g., factories or cars) that contribute to environmental degradation as well as the interaction of these units with each other and the environment. The system studies that are necessary to assess the trade-offs in such areas as materials choice (e.g., paper or plastic grocery bags, disposable or cloth diapers) are

1

difficult and frequently are hampered by lack of understanding of these interactions. Understanding the total system remains a daunting challenge.

This volume builds on earlier efforts of the National Academy of Engineering (NAE) in the area of technology and the environment.[1] It contains selected papers from the July 1997 Workshop on Industrial Ecology, Enabling Environmental Performance Improvement: The Role of Knowledge and Information Technology. The papers are presented in three sections. The first section explores the implications of information technologies for sustainable development and the legal context within which information and knowledge systems are evolving. The second section focuses on the areas where most of the path-breaking work is occurring—the individual corporation—and the information- and knowledge-sharing tools and techniques that are being developed in that arena. The third section provides examples of systems that are evolving in the relationships between corporations and society as a whole. Although the latter are still in development, they offer exciting potential for substantially improving the environmental efficiency of the economy.

This overview provides a context for the accompanying papers by discussing the role of information and knowledge systems in the evolving discipline of industrial ecology. It describes how companies are leveraging these systems to reap environmental benefits and how novel applications of information technologies are bridging the gap between industrial practice and society's interest in the environment and sustainable development. It concludes with suggestions on how to address some of the difficult issues related to "green" information and knowledge.

THE INFORMATION TECHNOLOGY REVOLUTION AND INDUSTRIAL ECOLOGY

Compared with the previous several decades, we now have a much better understanding of how human activities affect the environment. The vast majority of obvious environmental problems—caused by practices such as dumping trash and other waste material in open pits, disposing of wastewater in streams and rivers, and emitting emissions of pollutants into the atmosphere—are the result of what were once standard industrial practices. Steps to remedy these problems have focused on remediating specific sites and instituting compliance with, and enforcement of, end-of-pipe requirements and standards. Although adequate for their limited purposes of providing clean air, water, and land, these approaches increasingly are recognized as inadequate to deal with the more global perturbations of natural systems—climate change; loss of habitat and biodiversity; and depletion and degradation of soil, water, and atmospheric resources.

The knowledge base on which environmental decisions can be based is much broader and deeper than ever before. Ecology, which involves the study of the interactions among organisms and between organisms and their physical environment, continues to inform decision making across a wide range of applications, from agriculture and forestry to the design of artificial wetlands and the restoration of healthy ecosystems. Along with the other basic sciences, ecology will continue to improve the understanding of relationships between environmental concerns and human economic activities.

Some of these concerns are directly related (e.g., the link between chlorofluorocarbons and stratospheric ozone depletion). Solutions to such concerns (e.g., the Montreal Protocol and the development of environmentally friendly technologies and policies to speed their deployment) have tended to take into account industry's use of materials, energy, capital, labor, technology, and information, as well as the interaction of industrial systems with natural ecosystems. Industrial ecology is based on keeping track of the former and understanding the latter. Solutions based on industrial ecology include such approaches as designing goods and services in terms of their environmental life cycle so as to minimize environmental impacts and defining, assessing, and charting future technological directions to enable the achievement of sustainable development.

In industrial ecology, systems of production *and consumption* are considered as one. Therefore, solutions to environmental problems need to consider how production and consumption operate as a unit and interact with the large-scale environment. Yet much of environmental policy still focuses on manufacturing and production practices that often merely shift the problem elsewhere in the system. The more comprehensive view is critical when one considers, as Allenby (this volume) points out, the growth of the services sector. This sector, driven by information and knowledge acquisition and sharing, accounts for at least 60 percent of U.S. economic output and employment (U.S. Department of Commerce, 1996). The industries in this sector perform key economic and societal functions such as transportation, banking and finance, health care, public utilities, retail and wholesale trade, education, and entertainment. With the exception of transportation and utilities, these activities are not commonly associated with environmental impacts. Yet all consume energy and materials, and some, such as banking and financial institutions, indirectly influence the environment (e.g., through investment decisions). The service sector thus represents an untapped resource for environmental efficiency improvements. Service firms are well positioned to leverage their suppliers (upstream of operations) as well as their customers (downstream of operations) to effect systemic change (Richards and Kabjian, this volume). Their ability to do so can be enhanced by having better information upon which to base decisions.

To be successful, industrial ecology must adapt and incorporate technologies from any area that is found useful. Neither traditional environmental remediation—compliance or pollution control technology—nor "green" technologies alone are sufficient if environmental concerns are to be effectively mitigated. Information technology is a case in point. Never developed for environmental or "green" purposes, it nevertheless is creating new sectors of economic activity—most recently, electronic commerce—that is already changing the economics of industry. Freeman (1992) refers to the innovations in information and communications technology as technoeconomic revolutions—innovations that transform production and management throughout the economy.

Indeed, the current information and communications revolution is allowing pervasive changes to be made. The impacts of this revolution on the industrial metabolism of the economy and on industrial systems are being felt already, particularly in the monitoring and control of emissions; the use of energy and materials; the control of quality and inventory; and the improved control of manufacturing processes. Many of the energy-saving technologies and process changes that promote cleaner production depend on the incorporation of electronic sensors and monitors that provide input to control operations. System models of these processes often are complicated and their use requires online computers for proper implementation and compliance with many regulatory objectives.

Information and communications technologies also make possible improved quality and inventory control and help to reduce and eliminate defective or substandard products. This is not a result of the technologies themselves, but of a diffusion of a management philosophy associated with the technology. Pressures to reduce costs or to meet quality, design, performance, manufacturability, or environmental goals have been met by continuous improvements that are the result of the collective actions of all who are involved in the production or service function, or by users and customers. More recently, these improvements have been aided by the adoption of information technologies that help manage inventory and controls and capture and disseminate knowledge. Although the combined benefits of applying information technology with new management philosophies extend beyond a single plant to networks of plants, including outsourced activities, some of these practices may have negative environmental consequences. For example, just-in-time practices can lead to increased transportation (by truck, rail, and airplane) and associated increases in energy use and local air pollution.

Information and communications technologies also have resulted in fewer materials being used per unit product or function. For example, semiconductor technology uses vastly fewer materials and less energy than old vacuum-tube technology, and it is much more powerful. Similarly, on the materials front, there has been a reduction in metal consumption over the past 20 years (Sousa, 1992). Some of this reduction can be attributed to the information and communications revolution itself, which underlies improved product design systems. These systems use computer modeling to decrease reliance on prototypes. Information and communications technologies also have improved energy and material efficiencies because they have enabled innovations in new efficient manufacturing processes and the creation of new complex materials. The use of more-complex materials, however, has made recovery more difficult, and past experience shows that many previous environmental ills have resulted from the accumulation of materials in the environment. Hence, one might expect separation technologies to grow in importance as part of an overall environmental strategy.

The information and communications revolution is forging a far more integrated economy. At the same time, addressing environmental and sustainability concerns requires a multidimensional approach that is interwoven with the global economy and the planet's natural systems. Both factors, according to Allenby (this volume), are mutually reinforcing. This is because the concept of sustainability requires a global economy in long-term harmony with its supporting natural systems, which in turn will generate a far more robust economy—one that is more informationally dense, in which information is substituted for other inputs such as raw materials and energy. Citing economic trends in the information industry, Allenby shows that substitution of information for materials and energy has reduced the costs and use of these resources. He speculates that the demands for sustainability will increase the substitution of information for other inputs and postulates that sustainability itself may well be unattainable without such substitutions.

Information substitution, although an important contributor, will not, by itself, generate the ideal environment. As noted above, such substitutions are not without trade-offs, and "smart" policies will be needed. In the area of transportation, for example, there has been a merging of information and communications technologies in automobiles and traffic systems, including the development of so-called smart highways and vehicles to control traffic flow. The same has happened in air travel. Yet in neither case has the fundamental problem of reducing traffic been addressed. There are solutions, such as increasing ridership on public transportation. This may occur if significant improvements are made in transportation systems and if personal vehicle use is discouraged. Another alternative is to encourage people to work from home, telecommuting instead of traveling to work. Although such telework policies are beginning to appear in the workplace, gains from such practices can be offset easily by increases in other types of travel. For any of these approaches to be effective, the focus must be on addressing the problems of the total transportation system with a view toward minimizing the need for travel.

Hence, in many ways, information and communications technologies will continue to contribute positively to the environment in terms of reductions in materials and energy use. However, the final outcomes of such measures are likely to remain uncertain. Other areas in which application of the technology can contribute to environmental improvement include knowledge management—capturing information and knowledge so that past mistakes are not repeated (as discussed by Richards and Kabjian, this volume)—and knowledge creation. Legal barriers that are predicated on the traditional physical formats of knowledge, such as books, need to be addressed, according to Cohen and Martin (this volume). The current legal system is not well equipped to deal with "data mining" of publicly available information and to protect intellectual property rights in a world where access to information is easy and the information itself can be quickly reproduced.

At issue is data ownership. Is it the creator of the data or the individual who compiled them who has rightful ownership? Current intellectual property laws were not designed to protect and encourage the dissemination of compilations of factual information. They were designed to protect property. Creative expression and data do not fit well in either of these categories. Data are neither creative expressions like books, paintings, or sculptures, nor unique inventions. Database creators want protection the very moment that their data are gathered. In addition, databases are extremely dynamic and undergo constant change. As Cohen and Martin (this volume) point out, current patent and copyright laws are not suited to protect data or the compilation of data in a database. In the case of copyright, not only is current law ill-suited to the task, but it expressly bars protection of ideas, principles, and facts. In the case of patent laws, it can take years to process a patent application, and a clear definition of the unique invention is required.

Other laws, such as those related to trade secrecy and the tort of misappropriation, are equally ill-suited to protect the compilation of data. To address the common flaws intrinsic to the current intellectual property laws, Cohen and Martin suggest a two-phase approach that incorporates both property and liability. The first phase would provide a "blocking period" designed to give a certain amount of lead time for the database creator. During this period, a property rule would apply, and competitors would not be permitted to use or copy the new database without the database creator's consent. This initial blocking period would be followed by an automatic license. Absent some other agreement, the database creator would be obligated, at a minimum, to share the data with all secondcomers at rates established by a regulatory body composed of industry representatives and government officials. Under this approach, data creators would recover investments made during the compilation process, but the data would remain publicly accessible under fair and reasonable terms. This framework would serve society's interest in knowledge sharing, research, and development as well as data creators' legitimate interests in recouping development costs.

INFORMATION SYSTEMS WITHIN THE FIRM

The legal issues raised by Cohen and Martin are a product of our knowledge-driven society. Many experts in management believe that the manufacturing, service, and information sectors will be based on knowledge in the future, and business organizations will evolve into knowledge creators in many ways.

Drucker (1993) suggests that one of the most important challenges for every organization in the knowledge society is to build systematized practices for managing a self-transformation. Organizations have to abandon obsolete knowledge and learn to create new products and processes by improving ongoing activities and continually innovating in an organized way. Successful organizations of the future will have institutionalized the concept of growth based on knowledge creation and learning.

Three papers in this volume describe how private firms can develop information systems to better manage and create knowledge for environmental purposes. Richards and Kabjian point out that there are several opportunities to improve and apply environmental knowledge sharing, many of which cross traditional organizational boundaries. Such knowledge sharing may occur within a firm or involve the firm in collaborations with outside stakeholders that have interests in the company's environmental performance. Examples from DuPont (Carberry) and Rhône-Poulenc (Heptinstall) provide case studies of how individual firms in a heavily regulated sector—chemicals—are beginning to develop internal knowledge-sharing initiatives. Carberry shows how a vast array of information technologies such as e-mail, relational databases, CD-ROM, expert systems, Internet-based Web pages, teleconferencing, and videoconferencing is helping companies communicate environmental policies, exchange information about cleaner production technology, and report compliance data. In each instance, these new technologies provide for the rapid distribution or dissemination of environmental experiences, information, and knowledge that enhance technology transfer and enable companies to more effectively address compliance control and remediation. Hepinstall, on the other hand, discusses the challenges that firms face in implementing knowledge-sharing systems that share relevant environmental information internally within a company.

Graedel (this volume) and Ishii (this volume) explore another facet of the green technology challenge, namely, creating environmental knowledge that is of use to product designers. Graedel walks us through the design process, showing at what stages—from initial concept to final design—environmental knowledge can be useful. For example, when a product is in its conceptualization stage, he suggests addressing very basic environmentally related questions such as whether forbidden or highly regulated substances or materials will be required to manufacture the product and what the potential environmental impacts of the product throughout its life cycle, including recycling, might be. Some of the information needed to answer these questions may be located easily, but in other instances, the knowledge required may have to be created. Ishii illustrates one technique—the reverse fish-bone diagram—that designers can use to gain knowledge about parts and components of existing production. The purpose of undertaking such an exercise is to create knowledge that can be used in future designs to improve the recyclability of the product or family of products.

These examples show some steps of knowledge management and creation that a firm can take to improve its own environmental performance. Modern production operations, however, are nodes in an increasingly complex network of suppliers and distributors, which in turn require equally sophisticated knowledge systems if they are to be properly informed. Kleindorfer and Snir (this volume) explore environmental stewardship activities in this highly complex supply chain by focusing on how environmental information is gathered and used. They suggest that information technologies may help firms improve the environmental aspects of their products at three important levels: product and supply-chain design to minimize environmental impacts, ongoing waste minimization and risk mitigation after the product has been deployed, and diagnostic feedback from supply-chain participants to assess opportunities for new products and processes.

Heim (this volume), in turn, addresses how new software developments and the use of the Internet to distribute the software will allow small companies to model their manufacturing processes by accessing "plug-and-play" software components from various sources to develop manufacturing models of their operations. Whereas these models often can be used to optimize production, the technique of accessing such software and developing unique models is new. Similar applications may be developed that will help small manufacturers improve their environmental performance.

OPPORTUNITIES FOR COLLABORATION
AND NEW TECHNOLOGIES

Beyond the firm, environmental knowledge creation and management involve collaborations that are more complex. The complexity tends to be a huge obstacle that impedes the progress towards individuals involved working together effectively. Yet there is a critical need for collaborative work in the larger arena beyond the firm, and several collaborative arrangements have emerged. One is sector specific. In the for-profit world it takes the form of consortia of firms from a specific industrial sector working together on a particular problem. In the nonprofit sector, it takes the form of government agencies often forming task forces to work together on common issues. Another collaborative arrangement involves partnerships consisting of for-profit firms, private nonprofit interest groups, and the government that work on developing consensus on and solutions to issues of common interest. While the motivations that drive the two types of collaborations may differ, the challenges in both revolve around developing a common understanding of approaches to the problem at hand and establishing a standard terminology that all can work with.

Killgoar (this volume) makes the point that, from a private-sector perspective, the motivation for collaboration is to gain data, information, and knowledge. Using the automotive sector as an example, he describes the nontechnical, softer issues that arise in establishing and ensuring successful collaborations, such as building trust and developing common terminology. These issues, if successfully dealt with, can have enormous payback in development of new technologies. The challenge is to integrate information gleaned from these collaborative efforts into the operations of the constituent firms.

Government collaborations, on the other hand, are motivated by public-interest concerns such as getting information obtained by the government into wider circulation. The Environmental Data Exchange Network (EDEN) project, a collaborative effort of the U.S. Department of Defense (DOD), the U.S. Department of Energy, the U.S. Environmental Protection Agency (EPA), and the National Institute of Standards and Technology (NIST), is a case in point. These agencies have different types of related information from disparate sources and in different databases. According to Pitts and Fowler (this volume), EDEN seeks to provide a dynamic information system for accessing environmental data stored in diverse distributed databases. Like the collaborations in industry, the players involved in EDEN also had to agree on a framework of common approaches and a common terminology.

This collaboration also illustrates the innovative use of InfoSleuth™, a new software technology that uses intelligent software agents to provide uniform access to specific sets of information on geographically distributed environmental databases through standard Internet browsers. InfoSleuth™ itself was developed in a collaborative effort involving General Dynamics Information Systems (formerly Computing Devices International), NCR Corporation, Schlumberger, Raytheon Systems, Texas Instruments, TRW, and the DOD Clinical Business Area, and was partially supported by NIST. The motivation behind the development of InfoSleuth™ was to broaden the focus of current database research to produce a model that combines the semantic benefits of a structured database with the ease of publication and access of the World Wide Web.

Technologies like InfoSleuth™ will grow in importance as publicly available information changes the landscape of knowledge management and creation. Eagan, Wiese, and Liebl (this volume) describe Wisconsin's effort to develop an information system that will provide integrated environmental information about industrial facilities throughout the state via the Internet. Not only are socially responsible investor institutions on the quest for such information, but the public is also.

The extent to which information systems, mainly based on the Internet, support the development and distribution of environmentally relevant information and the potential power of this type of information distribution system usually is not well recognized, in part because of the newness of the medium. Already, however, global environmental information networks, complete with chat rooms and instant reporting of environmentally relevant events, are being developed (Knauer and Rickard, this volume). The Internet is unique in its ability to facilitate dialog.

Use of the Internet is enhanced further by effective organization of relevant information. Choucri (this volume) demonstrates how distributed knowledge-networking systems, such as the Global System for Sustainable Development (GSSD), can broaden the concept of merging knowledge from science with management prescriptions. GSSD is designed specifically for use in conjunction with Internet resources. Its knowledge base is organized as a hierarchical embedded system of entries about human activities and conditions; sustainability problems associated with human actions; current scientific and technological solutions; attendant economic, political, and regulatory solutions; and the broad range of evolving international actions and responses.

An example of how environmental information on the Internet is organized and used for broadcast and communication is provided at *http://www.scorecard.org*. This Internet site, established by the Environmental Defense (ED), pulls together Toxics Release Inventory data that companies report to the EPA and relates it to specific manufacturing sites on local- or national-scale maps. Knowledge is enhanced by linking information on specific chemicals to information on health and toxicity. By linking data and information, ED has put knowledge about emissions from specific industries and their potential harmful effects into the hands of individuals who may be affected. The existence of the Web site allows users to act on the information they find by, for example, communicating their concerns to responsible individuals in companies or to local regulators.

The implications of these developments for companies is that they have to be vigilant in providing accurate and meaningful information to the public. Knowledge, not just data, is particularly important because, with knowledge, the public can influence firms to change their behavior. Costs that reflect environmental performance and the availability of capital to address issues will have immediate and powerful impacts on firm behavior; similarly, public concerns about a firm or a nearby facility can create significant costs and even force curtailment of operations or closure of the facility. As the chemical industry learned—and responded to through the Responsible Care Program—public accountability is an increasingly powerful reality of corporate life. By reducing the level of false information, the credibility of all involved in environmental discussions will be heightened. If decision making is to be effective, it must be informed.

THE CHALLENGES AHEAD

As the papers in this volume show, an environmentally and economically efficient world will not necessarily be a simpler world; rather, it will be more complex and more informationally dense. There will be more, not less, demand for systems that can integrate information into knowledge across disparate spatial, temporal, and organizational scales.

These trends have at least three important public policy implications. First, there have to be incentives to generate environmentally relevant knowledge. From an industrial ecology perspective, such knowledge can impact the design of products, engineering or reengineering of ecological systems, communication with customers, understanding materials and energy flows, and research and development. Government support of academic research in this area can help identify new processes and techniques that enhance ecological objectives, articulate technical and management standards that reflect best strategic environmental approaches, and define criteria for determining environmental impacts and metrics of environmental performance. This narrow need within the realm of the industrial sector may seem trivial in the context of larger environmental issues of climate change and biodiversity but it is critical, particularly for the large and growing number of small and medium-sized manufacturers.

Second, there is a serious need to ensure that the environmental information is of high quality, not misinformation. The difficulty in this regard is that growth of scientific knowledge involves uncertainty. A useful safeguard might be a peer review process that continually assesses the validity of the information on which government agencies and private enterprises depend for decision making.

Finally, given the world's increasing technological sophistication and the close interaction between technological progress and environmental concerns, there is a need to develop a technologically and environmentally literate citizenry.

NOTES

[1]Over the past several years, the NAE Program on Technology and Sustainable Development has explored technology's impact on the environment through its role in production and consumption. Efforts in this area have focused on technology transfer (*Cross-Border Technology Transfer to Eliminate Ozone-Depleting Substances,* 1992); the relationship between science and environmental regulation and the effect of regulation on technological innovation (*Keeping Pace with Science and Engineering: Case Studies in Environmental Regulation,* 1993); industrial ecology and design for the environment (*The Greening of Industrial Ecosystems,* 1994); corporate environmental practice (*Corporate Environmental Practices: Climbing the Learning Curve,* 1994); The United States' and Japan's interest in industrial ecology (*Industrial Ecology: U.S./Japan Perspectives,* 1994); linkages between natural ecosystem conditions and engineering (*Engineering Within Ecological Constraints,* 1996); design and management of production and consumption systems for environmental quality (*The Industrial Green Game: Implications for Environmental Design and Management,* 1997); the diffusion patterns of environmentally critical technologies and their effect on the changing habitability of the planet (*Technological Trajectories and the Human Environment,* 1997); the impact of service industries on the environment (exploratory workshops held in December 1994 and June 1995); the examination of best or promising practices in several industrial sectors (*The Ecology of Industry,* 1998*),* an exploratory examination of metrics used by industry to gauge their performance and by ecologists to gauge ecosystem health *(Environmental Performance Metrics and Ecosystem Condition,* 1999); and a report on how the use of industrial environmental performance metrics may be improved (*Industrial Environmental Performance Metrics: Challenges and Opportunities,* 1999). Information on these publications is available online at <*www.nap.edu*>.

REFERENCES

Drucker, P.F. 1993. Post-Capitalist Society. New York: Harper Business.

Freeman, C. 1992. The Economics of Hope: Essays on Technical Change, Economic Growth and the Environment. London: Pinter.

Sousa, L.J. 1992. Toward a new materials paradigm. Minerals Issues (December). Washington, D.C.: U.S. Bureau of Mines.

U.S. Department of Commerce. 1996. Service Industries and Economic Performance. Washington, D.C.: U.S. Department of Commerce.

The Information Technology Revolution and Industrial Ecology

The Information Revolution and Sustainability
Mutually Reinforcing Dimensions of the Human Future

BRADEN R. ALLENBY

The automobile's evolution to a more environmentally efficient artifact shows how the information revolution and sustainability are mutually reinforcing. The automobile itself is a complex system. Its operation depends on several other systems such as roads and fuel delivery systems. These complex systems within systems require the generation and use of a wide range of information and feedback mechanisms. The examination of this evolution suggests wider implications for the information revolution and sustainability.

THE MODERN AUTOMOBILE: A PARABLE

Evolution of the Automobile

In the late 1960s automobiles were powered by what aficionados fondly called "Detroit iron": relatively crude but effective and large +400 in^3 V-8 engines. These "muscle cars" consumed enormous amounts of fuel, often getting less than 10 miles per gallon (mpg). Exhaust from these automobiles was untreated and contained high concentrations of hydrocarbon and nitrous oxide. But gas was cheap. Air was free. Environmental concerns were not yet widespread. Acceleration was great. Then came Earth Day in 1970 and the energy crises of the early 1970s. Pollution control equipment was superimposed on existing engine designs. Demand for improved gas mileage increased. The passage of the Energy Policy and Conservation Act in 1975 (15 U.S.C. §§ 2001 et seq.) established corporate average fuel economy requirements for new automobiles. Accordingly, in the early and mid-1970s, average engine size, engine efficiency, and performance dropped.

15

Yet the decrease in automotive performance, measured along almost any parameter, was temporary. The average size of the engine in passenger cars for sedans with four-, six-, and eight-cylinder engines fell from approximately 290 in^3 in 1975 to about 180 in^3 in 1992 and stayed small.[1] Horsepower (hp) also dipped over this period. However, the ratio of hp to engine displacement increased significantly, from about 0.5 hp/in^3 in 1975 to over 0.8 in 1991, indicating more efficient operation. Moreover, the fuel economy of the average new car improved significantly, from 15.8 to 27.8 mpg between 1975 and 1991. At the same time, absolute performance of the product increased: Acceleration from 0 to 60 miles per hour went from 14 seconds in 1973 to about 12 seconds in 1991 (Graedel and Allenby, 1997; National Research Council, 1992). The modern automobile unquestionably provides more performance per unit resource (in this case, gasoline). Moreover, today's automobile is considerably safer, handles better, lasts longer, and offers far more amenities, such as advanced sound systems, onboard diagnostics, and climate control systems. Impressively, these gains have been matched by similar increases in environmental efficiency: Since controls were introduced in 1968, volatile organic chemicals and carbon monoxide emissions per vehicle have been reduced by some 96 percent, and, since the imposition of nitrous oxide controls in 1972, emissions of those species have been reduced by over 75 percent (MacKenzie, 1994).

In short, over the past two and a half decades, one of the principal and defining artifacts of the modern industrial economy has undergone an almost revolutionary change. It has improved substantially its environmental performance on a per unit basis; it is a far safer and more desirable product; and it has significantly enhanced not only its performance, but the efficiency with which it generates that performance.

Information Technologies and the Automobile

The performance of the modern automobile reflects a number of incremental improvements: reductions in vehicle weight, better aerodynamic design, reductions in tire rolling resistance, reduction in friction losses, new catalytic systems, and more efficient engines and drivetrains. But there is one common theme underlying the evolution of the modern automobile: It has become a much more complex system, with a far higher information content than its predecessors, and it is increasingly linked to its external environment, becoming a subsystem in a yet more complex automotive transportation system (Figure 1).

Internally, subsystems in older cars were linked mechanically. Systems in new cars are linked by sensors feeding into multiple computers. Whereas older cars had minimal electronics, newer ones have substantial systems that need to be integrated both physically and functionally. The number of cables and wiring harnesses required by the modern automobile has increased to such an extent that routing them through the vehicle becomes a design problem in itself (Thompson,

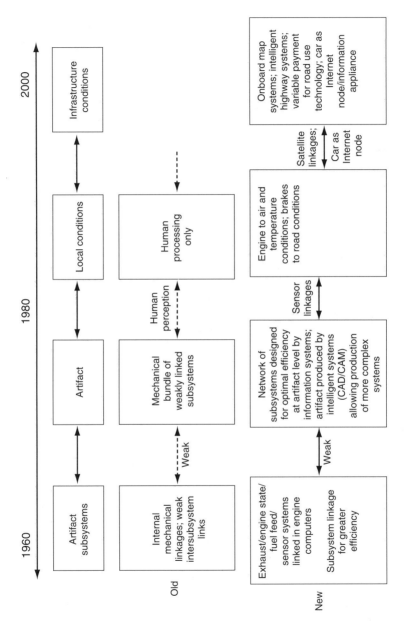

FIGURE 1 Technological evolution of the modern automobile. Note: CAD, computer-aided design; CAM, computer-aided manufacturing.

1996). For example, "[d]oors . . . can barely be made to open and close properly, what with wires for window controls, locks, outside mirror controls, and other switches and lights." Automobile manufacturers now request complete engine management systems to balance performance, emissions, fuel consumption, and operating conditions (e.g., cold starts, stop-and-go conditions). Such systems typically include control units, sensor systems, and actuators and use complex information topologies to maintain optimal system operation (Krebs, 1993). The modern automobile, therefore, reflects a much more complex engineered system in which sophisticated multiplexed microcontrollers have become a necessary component (*The Economist*, 1994b).

Producing a more complex system requires, in turn, more sophisticated design tools and manufacturing technologies. For example, lightweighting (reducing the weight of vehicles through better design and material substitution) has been a major contributor to improving environmental performance. Such designs require precision manufacturing and become a far more information-intensive activity. Germany's Audi, for example, believes that only the advent of supercomputing technologies provided the necessary processing power to design and model the performance of the complicated lighter components that have permitted them to lightweight their product. Difficult design problems are resolved using virtual reality design processes. Indeed, powerful computer-aided design systems can replace, with a click of the computer mouse, hours of laborious work done on thousands of drawing boards (*The Economist*, 1994a). In fact, Boeing dispensed with building a physical model of its latest aircraft, the 777. Instead, the aircraft was created entirely within a distributed computational system using sophisticated simulation software.

As it is with the artifact, so it is with the built infrastructure system within which it functions. In older cars, virtually the only information link between the automobile and the external environment was the driver. Today, sensor systems monitor exhaust systems, the oxygen content of airflows, and road conditions. Newer systems map the car's geographical position, provide up-to-date road conditions and optimal real-time routing options, and pay tolls electronically without the need to stop. Technologies already exist that will permit ongoing communication between road networks and automobiles. This would, in essence, integrate the automotive built infrastructure, the automobile, and the driver as one automotive transportation system, which in turn can be optimized to provide real-time efficiency by, for example, using time-of-day and location sensitive automatic roadway pricing (Jurgen, 1995). In fact, at the Cyberhome exhibit in San Francisco in 1997, Mercedes-Benz displayed a multimedia concept car, linked to the Internet with speech recognition capabilities, a voice-controlled browser, global positioning system capability, and its own internal local-area network. The car essentially is conceptualized as an information appliance. In discussing the vehicle, Lewis (1997) in *Scientific American*, claimed, perhaps optimistically, that "The society of Web cars will be able to get themselves out of traffic jams, avoid bad weather and keep their inhabitants well informed and entertained. With such a huge potential market waiting for manufacturers, Web cars are inevitable. Exactly what form they will take remains to be seen." Possibilities include speech recognition capability which will allow, for example, dictation of letters and e-mail (Floren, 1997).

This evolution of a more environmentally and economically efficient auto-mobile provides an analogy for at least some of the characteristics of a more sustainable economy. To the extent that the analogy is valid, it suggests that such an economy will be more, not less, complex and, concomitantly, far more infor-mationally dense. Information generation (through, for example, appropriate sys-tems of sensors), the evolution of more complex feedback systems, and tighter linking of previously disconnected subsystems through new information links (e.g., intelligent cars on intelligent roadway systems) will support a fundamental pattern: the substitution of data and knowledge information for other, less envi-ronmentally appropriate inputs into economic activity.

THE INFORMATION REVOLUTION AND
SUSTAINABILITY IN CONTEXT

The parable of the automobile suggests a fundamental coevolution of the Information Revolution and sustainability: that greater environmental efficiency will require, as an enabling capability, the Information Revolution, and that the latter, in turn, will be strongly encouraged by the need for greater environmental efficiency. Here, of course, environmental efficiency is not taken as the usual green technology but, rather, as the reengineering of the Industrial Revolution suggested by the nascent, integrative science of industrial ecology. Table 1 shows the difference between green technology and environmentally preferable technol-ogy systems. After all, the goal is not control of local perturbations or acute human risk, but sustainability.

If sustainability is a state that emerges only at the level of a global human ecology, which is a highly defensible hypothesis, then subsystem sustainability cannot be defined except in terms of relationship to global systems, and certainly the knowledge or wisdom to know what that means is not on hand. This is particularly true because there are probably many sustainable states. Therefore, it becomes a value judgment and a function of the ability of human institutions to adapt to an environmentally constrained world that will determine toward which state humanity moves (Cohen, 1995; Allenby, 1998). Indeed, industrial ecology is intended to provide the science and technology base for understanding what sustainability actually might mean (Graedel and Allenby, 1995; IEEE, 1995; Allenby, 1997).

This does not mean, however, that humanity must drift: It is entirely possible to define environmental efficiency as providing equal or greater units of the quality of life while reducing the resultant summed environmental impacts

TABLE 1 Green Technology versus Environmentally Preferable Technology

Technology Designation	Philosophy	Example	Origin	Market Structure	Endpoint
Green technology	The government can mandate everything we need	Scrubbers, water treatment plants, pollution prevention	Localized effects, command and control, end of pipe	Central mandate and control	Reduce local human risks
Environmentally preferable technology systems	Evolution of complex systems within appropriate boundary conditions	Central digital servers providing video/music/multimedia on demand; energy- and water-efficient dishwasher	Technological change, implementation of industrial ecology theory and design for environment	Internalization of externalities; free-market function within boundary conditions	Sustainability

integrated across the life cycle of the process, product, service, or operation involved. Quantifying these impacts in specific instances is quite difficult, but the principle is understood easily, as the case of the automobile discussed above.

It is also important *ab initio* to recognize that the information infrastructure that supports information services of all kinds, from sensor systems to tele-communications, itself is not without environmental cost. For example, part of the cost of the increased efficiency of the automobile is the consumption by that sector of some 12 percent of the printed wiring boards produced in the United States annually (compared with about 39 percent that go into computational devices) (MCC, 1994). More broadly, the production of electronics components and subassemblies, their energy consumption over their life cycles, and the end-of-life treatment of electronics products sometimes generate significant environmental impacts of different types. Chip production, for example, consumes substantial energy and water resources (see Table 2), amounting to about half of the cost of semiconductor manufacturing (MCC, 1994). Moreover, the rapid pace of the technological evolution of electronics and the highly competitive international markets for these goods result in rapid product obsolescence and thus substantial generation of waste materials as old products are discarded. Problematic issues include the leaching of heavy metals from solder and leaded glass used in displays and the sheer volume of discarded electronics items (MCC, 1993).

Realization of the full environmental and economic benefits of the substitution of information for other economic inputs, therefore, requires that the environmental impacts of electronics products across their life cycle be minimized. In the electronics industry, this is accomplished by using design for environment (DFE) methodologies and tools, which have evolved rapidly.[2] And, although

TABLE 2 Semiconductor Manufacturing Energy Consumption

Input	Cost (thousands of dollars)	Percentage of Manufacturing Cost
Central plant	3,720	7.1
Building structure	3,838	7.3
Ultrapure water	539	1.0
Chem services	479	0.9
Gas services	247	0.5
Electricity	21,890	41.5
Water	1,423	2.7
Natural gas	2,481	4.7
Custodial (cleaning service)	1,953	3.7
Gounds landscaping	102	0.2
Trash	158	0.3
Hazardous waste management	774	1.5
Salary/benefits	15,105	28.7

SOURCE: Based on Lando, 1996.

DFE has not yet been fully understood and adopted by any firm, it is increasingly being integrated into many of the concurrent engineering systems of leading electronics firms around the world. Moreover, governments in Europe and Japan are actively exploring ways to minimize the environmental impacts of electronics products, including imposing postconsumer product takeback requirements on manufacturers of consumer electronics. Thus, the environmental impacts of the platforms by which information services are provided are at last beginning to be addressed.

EMERGENCE OF THE INFORMATION INDUSTRY

Several salient points are now apparent about the Information Revolution. The first is the emergence of an information industry from previously disparate sectors (Figure 2), which is seen as creating an industry sector with four components: information content, information servers, information networks, and information appliances. Note that, as suggested by the automobile example, these should not be taken as stand-alone sectors. Rather, they are functions, which are increasingly embedded throughout the economy, from manufacturing and agricultural operations to products of all kinds; to service systems such as transportation and energy; to end-use applications such as the Internet, intranets, computers, telephone systems, and televisions.

The second point is the pervasive use of information technologies within the economy and their vital role in enabling the development of a robust service sector, which has grown to dominate developed economies, even manufacturing-oriented ones such as Japan's (Figure 3). Although there is considerable ambiguity about the term "services," the service sector includes transportation, communications, utilities, and sanitary services. Table 3 provides a breakdown of the Standard Industrial Classification divisions and the major groups representing services. It also shows that the service industry groups are far broader than the information industry, however the latter is defined.

There is a more subtle truth in the intuitive linkage between the information industry and the service economy, however. The Industrial Revolution could not have occurred without concomitant breakthroughs in agriculture, manufacturing, energy, and transportation infrastructures. It also was based on its own information revolution: The printing press and nascent media were critical elements in the diffusion of knowledge, particularly scientific and technological, without which the rapid growth of industrialization and supporting infrastructures could not have occurred. The Industrial Revolution, which fundamentally represents the meeting of human wants and needs—providing a demanded level of quality of life—through increased material wealth, was fueled by its own information revolution. Similarly, now, the current Information Revolution offers the promise of providing enhanced quality of life through services, not simply material acquisition. The challenge then becomes substituting services for material products, and dematerializing services,

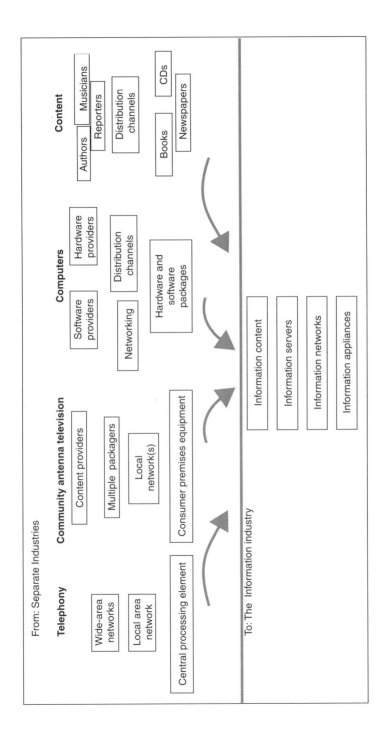

FIGURE 2 Emergence of the information industry. SOURCE: Adapted from Lando, 1996.

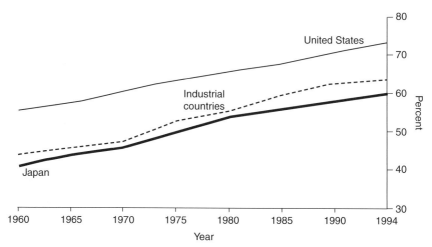

FIGURE 3 Service-sector employment in developed economies. SOURCE: Adapted from IMF, 1997, p. 47.

TABLE 3 Major Standard Industrial Classification Divisions and Groups Representing the Service Sectors

Code	Classification
Production Sectors	
Division A	**Agriculture, forestry, and fishing**
Division B	**Mining**
Division C	**Construction**
Division D	**Manufacturing**
Service Sectors	
Division E	**Transportation, communications, electric, gas, and sanitary services**
Major Group 40	Railroad transportation
Major Group 41	Local and suburban transit and interurban highway passenger transportation
Major Group 42	Motor freight transportation and warehousing
Major Group 43	U.S. Postal Service
Major Group 44	Water transportation
Major Group 45	Transportation by air
Major Group 46	Pipelines, except natural gas
Major Group 47	Transportation services
Major Group 48	Communications
Major Group 49	Electric, gas, and sanitary services
Division F	**Wholesale trade**
Major Group 50	Wholesale trade—durable goods
Major Group 51	Wholesale trade—nondurable goods

TABLE 3 *continued*

Code	Classification
Division G	**Retail trade**
Major Group 52	Building materials, hardware, garden supply, and mobile homes
Major Group 53	General merchandise stores
Major Group 54	Food stores
Major Group 55	Automotive dealers and gasoline service stations
Major Group 56	Apparel and accessory stores
Major Group 57	Home furniture, finishings, and equipment stores
Major Group 58	Eating and drinking places
Major Group 59	Miscellaneous retail
Division H	**Finance, insurance, and real estate**
Major Group 60	Depository institutions
Major Group 61	Nondepository credit institutions
Major Group 62	Security and commodity brokers, dealers, exchanges, and services
Major Group 63	Insurance carriers
Major Group 64	Insurance agents, brokers, and services
Major Group 65	Real estate
Major Group 67	Holding and other investment offices
Division I	**Services**
Major Group 70	Hotels, rooming houses, camps, and other lodging places
Major Group 72	Personal services
Major Group 73	Business services
Major Group 75	Automotive repair, services, and parking
Major Group 76	Miscellaneous repair services
Major Group 78	Motion pictures
Major Group 79	Amusement and recreation services
Major Group 80	Health services
Major Group 81	Legal services
Major Group 82	Educational services
Major Group 83	Social services
Major Group 84	Museums, art galleries, and botanical and zoological gardens
Major Group 86	Membership organizations
Major Group 87	Engineering, accounting, research, management, and related services
Major Group 88	Private households
Major Group 89	Miscellaneous services
Division J	**Public administration**
Major Group 91	Executive, legislative, and general government, except finance
Major Group 92	Justice, public order, and safety
Major Group 94	Administration of human resource programs
Major Group 95	Administration of environmental quality and housing programs
Major Group 96	Administration of economic programs
Major Group 97	National security and international affairs
Division K	**Nonclassifiable establishments**

SOURCE: Bureau of the Census, 1996.

in a large part through the substitution of information technology and intellectual capital (increasingly embedded in software) for material and energy input. There are some indications that this is, in fact, occurring: Some 80 percent of all information technology in the United States, for example, is purchased and used by the service sector (Rejeski, 1997). Nonetheless, our understanding of, and ability to assess, this process is quite limited at this preliminary juncture, and it would be premature to draw any firm conclusions.

For one example, the substitution of information for physical inputs is liable to be quite subtle in some cases. Consider, for instance, an illustrative and anecdotal example drawn from a nonservice sector, agriculture. It would appear at first glance to have little, if anything, to do with information markets. However, on closer examination, one finds that some firms are beginning to offer pest management services, in lieu of simply selling biocides. Such services tend to rely on more complex mixtures of technologies, require more knowledge of pests and their habits (not to mention their genetics), and make less use of biocides than in existing practices (Benbrook et al., 1996). They are, in short, substituting complexity and information for material consumption in the agricultural process by offering a service rather than a product.

More fundamentally, Monsanto's CEO, Robert Shapiro, explains another way in which information is being substituted for material inputs such as pesticides (and the concomitant "inert ingredients") and the energy embedded in them and used to apply them (Magretta, 1997):

> We don't have 100 years [to figure out how to avoid ecological catastrophe or food shortages]; at best, we have decades. In that time frame, I know of only two viable [mitigation technology] candidates: biotechnology and information technology. I'm treating them as though they're separate, but biotechnology is really a subset of information technology because it is about DNA-encoded information.
>
> Using information is one of the ways to increase productivity without abusing nature. . . . Sustainability and development might be compatible if you could create value and satisfy people's needs by increasing the information component of what's produced and diminishing the amount of stuff. . . .
>
> I offer a prediction: The early twenty-first century is going to see a struggle between information technology and biotechnology on the one hand and environmental degradation on the other. Information technology is going to be our most powerful tool. It will let us miniaturize things, avoid waste, and produce more value without producing and processing more stuff. The substitution of information for stuff is essential to sustainability.

Another example, still in its infancy, is the provision of textual material, data, and software upgrades through the Internet and via electronic mail. This is a broad trend, and a few examples should suffice to illustrate it. Many journals now accept, if they do not require, submission of manuscripts in electronic form. Data appendices, which previously were available largely in hard copy, now are routinely available online from central databases. Indeed, it is doubtful that data-intensive efforts such as the Human Genome Project could have been undertaken without such information access. Several journals are completely online now, and some publishers have imprints dedicated to electronic publishing. A topical example can be taken from AT&T. Only a few years ago, AT&T printed some 100,000 copies of its annual environmental report, because hard copy was all that was available; in 1996 the number of reports printed dropped to 10 percent because the entire report was available on the Internet. Sun Microsystems has gone to a completely electronic format for its annual environmental report (Craig, 1997). Software upgrades are increasingly provided over the Internet, reducing the need to send thousands of floppy disks or CD-ROMs through the mail (and the concomitant environmental impacts associated with the manufacture and distribution of these artifacts).

RELEVANT TRENDS IN THE INFORMATION INDUSTRY

Several trends in the information industry stand out and tend to suggest continued substitution of information for other inputs into the economy. The dramatic and continued increase in processing capacity of microprocessors and in memory per chip (Figure 4) are well known. What is equally important, however,

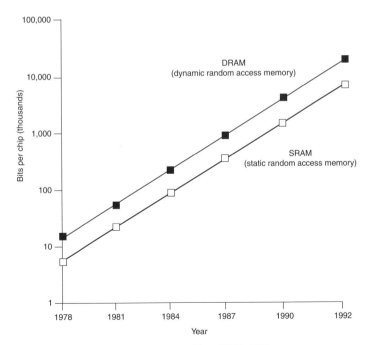

FIGURE 4 Increased capacity of memory chips, 1978–1993.

is that the rate of technological evolution in the information industry is driving order-of-magnitude improvement in virtually all relevant technologies, from optical transmission to information storage, to signal compression, to efficient spectrum use, to software (Table 4). Equally important, the costs of information manipulation continue to fall, with exponential improvements in such critical metrics as flops per dollar (flops are floating-point operations, a measure of computational performance) (Figure 5). In fact, digital computation trends show that, on a per million instructions per second (MIPS) basis, digital computation technology is exponentially dematerializing, shrinking in volume, gaining in power efficiency, and gaining in economic efficiency (Figure 6). Because the first three parameters arguably capture many sources of environmental impact and the latter obviously captures economic impact, it is hard to argue that significant simultaneous gains in environmental and economic efficiency are not at the core of historical and current performance of the information industry.

Accordingly, consumption of information platforms is increasing, in many cases exponentially. This is true for information "pipes," as shown, for example, by the increase in voice paths over the Pacific (Figure 7), and by the number of new subscribers to information services such as cellular telephones (Figure 8). Although the data are hard to evaluate, this is apparently leading to substantial growth in information stocks, most particularly in new electronic media as opposed to traditional hard media such as books (Figure 9). Although the data are hard to evaluate, this is apparently leading to substantial growth in information stocks, particularly in new electronic media as opposed to traditional hard media, such as books (Figure 10 presents such data from Japan). Information consumption appears to be growing more rapidly than either gross national product (GNP) or population, at least in some developed countries (Figure 10). Overall, the impression is of an expanding information industry sector with rapidly evolving technology and falling costs per unit of performance.

Material and energy trends are more complex. Substitution effects—new materials for old, for example—on all scales, from the economy itself to sectors to specific applications, and their tangled relationship with technological evolu-

TABLE 4 Technological Advances in the Information Industry

Technology	Technology Trend
Silicon chip	Multi-2 in speed every 18–24 months
Optics	Multi-2 in speed every 3.5 years
Storage	Reduction of cost by half every 2 years
Energy	Distributed generation
Compression	30:1 in space in past 5 years
Spectrum	Better reuse, higher availability
Software	Portable operating systems, middleware, distributed systems

SOURCE: Adapted from Lando, 1996.

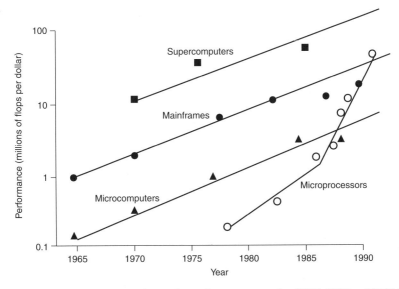

FIGURE 5 Central processing unit performance trends, 1965–1990. SOURCE: Reprinted from *Science*. Buzbee, B. 1993. Workstation clusters rise and shine. *Science* 261(5123):852–853.

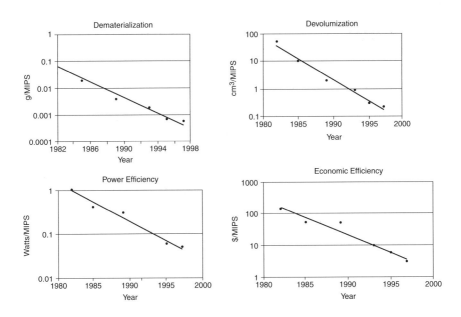

FIGURE 6 Eco-efficiency of digital computation. SOURCE: Adapted from Lando, 1996.

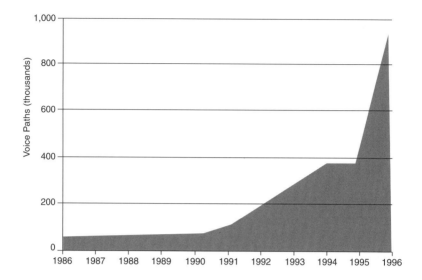

FIGURE 7 Transpacific voice paths, 1986–1996. SOURCE: Adapted from Pacific Economic Cooperation Council, 1996, p. 82.

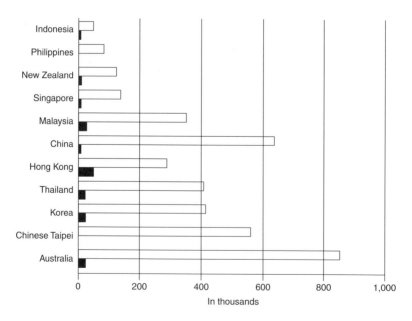

FIGURE 8 Growth in Asian cellular subscribers, 1988 versus 1994. SOURCE: Adapted from Pacific Economic Cooperation Council, 1996, p. 83.

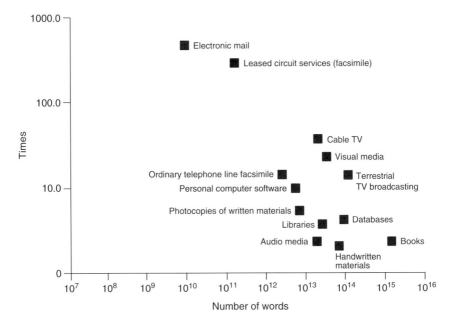

FIGURE 9 Growth of information and communication equipment stocks in Japan, 1983–1993. SOURCE: Japan Ministry of Posts and Telecommunications, 1995, p. 16.

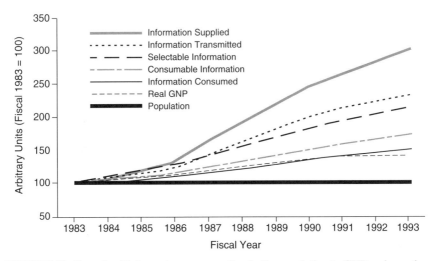

FIGURE 10 Growth of information consumption in Japan relative to GNP and population, 1983–1993. SOURCE: Adapted from Japan Ministry of Posts and Telecommunications, 1995, p. 13.

tion, are difficult to assess in general terms. Moreover, a thicket of subsidies, direct (e.g., depletion allowances) and indirect (e.g., subsidized transportation for virgin materials), is common in energy and material sectors, further complicating analyses. Nonetheless, some trends appear to be fairly robust.

Regarding energy, it is apparent that consumption will continue to increase strongly over the next several decades as a result of continuing economic development, particularly in Asia. Growth rates will be higher than population growth (i.e., higher per capita energy consumption with development) but less than gross domestic product (GDP) growth (i.e., continuing the decoupling of energy consumption from economic growth rates) (EIA, 1996). The shift from primary fuels to electricity, which has characterized the evolution of developed economies, will continue on a global basis. The U.S. Energy Information Administration (EIA) anticipates, however, that existing energy resources and technologies are adequate to meet this demand and thus concludes that "the real cost of energy need not escalate significantly over the projection period [to 2015]" (EIA, 1996). It is not clear, however, whether this projection takes into account the massive investment required for development in Latin America and Asia and for supporting rapid technological evolution (and the associated equipment, facility, and infrastructure replacement) in developed economies. These demands for capital could raise its cost (interest rates) significantly and thus make the building of new energy infrastructures very expensive, thereby making the provision of energy more expensive.

The materials picture is somewhat more complex. Although it is generally believed by some that the materials efficiency (material used per unit of GDP or, more broadly but less easily measured, per unit of quality of life) of developed economies is improving, data are ambiguous. Thus, for example, the survey of material use patterns in the United States by Wernick et al. (1996) leads them to conclude that, although there are theoretical reasons to believe that dematerialization of economic activity will proceed, current trends are unclear:

> With regard to primary materials, summary ratios of the weight of materials used to economic product appear to be decreasing due to materials substitution, efficiencies, and other economic factors. The tendency is to use more scientifically selected and often artificially structured materials. These may be lighter, though not necessarily smaller. The value added clearly rises with the choice of material, but so may aggregate use.
>
> With regard to industry, encouraging examples of more efficient materials use exist in many sectors, functions, and products. Firms search for opportunities to economize on materials, just as they seek to economize on energy, labor, land, and other factors of production. However, the taste for complexity, which often meshes with higher performance, may intensify other environmental problems, even as the bulk issues lessen.
>
> With regard to consumers, we profess one thing (that less is more) and often do another (buy, accrete, and expand). No significant signs of net dematerialization at the level of the consumer or saturation of individual materials wants is evident.

With regard to wastes, recent, though spotty, data suggest that the onset of waste reduction and the rapidity with which some gains have been realized as well as the use of international comparisons indicate that very substantial further reductions can take place.

Cost trends are difficult to generalize but, in general, waste disposal costs, very roughly tied to toxicity, are increasing, as are prices for primary commodities, which from 1970 to 1997 have trended upward, although with significant variation (IMF, 1997). Even if prices for fuel and nonfuel materials are assumed to remain stable (perhaps as a result of substitution and technological evolution with concomitant increases in efficiency of process and use of materials), the cost differential favoring substitution of information for other inputs should continue to grow because of the steep continuing decline in costs of the former.

Significantly, these commodity price trends are reflected in a fundamental shift in the values accorded to firms by the financial markets. Microsoft, for example, with minimal physical assets (e.g., its campus in Redmond, Washington) but abundant intellectual capital (its people), has a market capitalization greater than Ford, General Motors, and Chrysler taken together, despite their huge asset bases. As Walter Wriston (1997), former chairman and CEO of Citicorp/Citibank and chairman of the Economic Policy Advisory Board in the Reagan administration, notes:

> The pursuit of wealth is now largely the pursuit of information and its application to the means of production. The rules, customs, skills, and talents necessary to uncover, capture, produce, preserve, and exploit information are now humankind's most important. The competition for the best information has replaced the competition for the best farmland or coal fields. . . . The new economic powerhouses are masters not of huge material resources, but of ideas and technology. The way the market values companies is instructive: it now places a higher value on intellectual capital than on hard assets like bricks and mortar.

More generally, knowledge as a critical factor of production, like capital, labor, and raw materials, is increasingly recognized by prominent economists such as Paul Romer (Kurtzman, 1997) and industrial theorists such as Arie de Geus (1997), known for his innovative introduction of scenario planning techniques at Royal Dutch Shell. The implications of this shift to knowledge as a critical input to the firm are profound; for the economy as a whole, it has been suggested that efficiency in a knowledge economy requires equity (Chichilnisky, 1996).

These data are all suggestive, but they are neither systematic nor comprehensive. What would be desirable is a rigorous way of determining whether the substitution of information and intellectual capital for other economic inputs is actually occurring and, if so, to quantify and track this trend over time. Conceptually, one thus wishes to measure the information density of the economy, which, like many useful and obvious activities, is easier said than done. Although a possible theoretical approach is sketched below, much further effort is required to develop a workable, quantitative measure.

INFORMATION DENSITY OF AN ECONOMY

Concept

One way of thinking about a more complex economy is in terms of its information density, especially if greater information density is likely to be a necessary, if not a sufficient, requirement for greater economic and environmental efficiency. It is, of course, also apparent that a more complex economy, in itself, does not guarantee such efficiency: One could simply devise more numerous and more complex ways of producing more onerous environmental impacts. Thus, although this section focuses on the concept of information density, a second step needed to link the Information Revolution rigorously to sustainability is to quantify and understand the correlations and, if feasible, the causal linkages between increasing information density and greater environmental efficiency. The latter step will require substantial research and conceptual development.

Moreover, it is also apparent that more information, by itself, may not equate directly to greater knowledge—"useful information"—and thus enhanced quality of life: junk faxes and e-mail bedevil many of us; and the Internet, although useful, is also information pollution raised to an art form, a postmodernist information Superfund site in the making. Much of this overabundance of information is probably a reflection of the youthful exuberance of the fundamental shift to a knowledge economy and temporary price distortions in a rapidly changing market (e.g., "free" Internet information), and it is at least probable that much information pollution will disappear as market forces begin to shape a competitive information and knowledge economy. It must be recognized, after all, that there is some value in information redundancy in a complex system. More fundamentally, someone is consuming the films and videos and using up telecommunications and Internet transmission capacity almost as soon as it is installed: Anyone with teenagers can attest that one person's noise is another person's rock and roll (or, somewhat more rigorously, the transition from information to knowledge is heavily contextual and subjective).

Conceptually, the information density of an economy is somewhat analogous to physical density and thus can be given by a formula based on that for physical density (volume divided by mass):

$$D_i = \frac{V_i}{E_a}$$

where D_i is the information density of the economy in bits per dollar, V_i is the volume of information in the economy in bits, and E_a is the economic activity in the economy measured in dollars. Unlike physical density measures, however, information density involves a time domain because both stocks and flows of information, and economic activity, are involved. Thus, information density might have to be an averaged figure (perhaps over a year to match GDP data). Economic activity in the aggregate is relatively easy to measure (but, even here, appropriate caution must be exercised: For example, natural resource accounting methods are primitive and seldom used). Moreover, it would be preferable to measure quality of life per unit of information rather than simple economic activity, but metrics for such subjective dimensions of the human experience are hard to validate, and the correlations between them and economic indicators are not well understood. Even now, for example, it is proving difficult in practice to evaluate increases in quality of life due to improvements of existing artifacts where such increases are not reflected in price changes.

Quantifying the volume of information in the economy is the problematic step. Conceptually, one way of defining V_i is as the number of bits (or bit equivalents, for analog systems) communicated through the communications networks of the economy or consumed in a given period, including data, voice, and video (volume telecommunicated, or V_t); the number of bits consumed in the economy through, for example, listening to music or watching a videotape (volume consumed, or V_c); the number of bits generated within artifacts, such as cars, airplanes, and coffee makers (volume in artifacts, or V_a); the number of bits generated within facilities and infrastructure, such as manufacturing facilities, administrative buildings, retail outlets, fast-food establishments, and the like as part of their operations (volume in facilities, or V_f); the number of bits published in other than electronic media and not duplicated there (V_p); and the information content in all other residual uses (V_r).

$$V_i = V_t + V_c + V_a + V_f + V_p + V_r$$

Defining these amounts will be difficult, especially because the terms are not orthogonal; some of the information consumed, for example, has been transmitted previously.

Some of these terms, such as the volume telecommunicated, can in theory be estimated from existing data that already have been collected, although the practical problems involved in actually doing so are substantial. Some terms, for example, might be estimated by multiplying the information capability of the relevant universe by the number of times that capability is accessed. Thus, an estimate of bits consumed could be derived by multiplying the number of bits stored in various media by the number of times that the media unit is accessed and then multiplying that figure by the audience per time accessed (the Star Wars movie, for example, might make a significant contribution to the amount). The number of bits generated within artifacts also might be relatively easy to estimate, based on the capability of the chips embedded in the devices and the access rate. A similar process, adjusted for double counting, might be applied to bits generated within facilities and infrastructure. A figure for bits published in nonelectronic media forms should be relatively easy to estimate, although double counting might be a problem here as well.

Overall, however, these terms should be considered as illustrative of the concept; as discussed below, actually measuring the information density of the economy will be a complex and intensive task, not easily accomplished without a focused effort involving experts in a number of fields.

The easiest way to measure E_a is in terms of dollars of economic activity, but this has some measurement difficulties as well. Many of these, such as costing noneconomic but productive work such as housework or raising children, are familiar to economists, however, and reasonable valuation methods can be used to impute proper figures. More fundamentally, in keeping with the industrial ecology approach, which encompasses both economic activity and associated externalities, E_a should be considered as the sum of measured economic activity, or E_m, and externalities, or E_x:

$$E_a = E_m + E_x$$

The difficulties of quantifying externalities, a category that includes but goes beyond many of the proposed green accounting systems, are substantial, but not impossible if absolute precision is not required. This provides the full equation

$$D_i = \frac{V_t + V_c + V_a + V_f + V_p + V_r}{E_m + E_x}$$

Estimating this indicator for the period from the beginning of the Industrial Revolution until the present could be an interesting way of determining whether the vaunted Information Revolution is, in fact, occurring. It also might be one indicator, albeit insufficient without others, for progress toward a more environmentally and economically efficient economy. (An interesting question for further research is whether V_i can be broadly defined in such a way as to generate a quantitative measure of the complexity of society as a whole, rather than just the economy.)

Metrics

The theoretical appeal of a robust metric for information density is considerable, but actually developing such a measure will be quite challenging. An obvious initial step, for example, is to turn to existing economic databases and measures, which have the significant advantages of being traditional, relatively standardized, and globally ubiquitous. Could, therefore, a very rough estimate of the information density be obtained by using economic data alone; using summed data on consumption in information sectors such as books, cable television, telephone, CDs; and comparing the ratio of that sum to economic activity as a whole?

Although it would fairly easily yield a result, this approach could be seriously misleading, as demonstrated by Table 5, which uses Bureau of the Census data to track information industry economic performance as a percentage of the GDP for selected years from 1980 to 1994 (this data format extends back only to

TABLE 5 Traditional Information Sector Share of GDP

		Share of GDP, 1980 and 1987–1994 (billions of current dollars, rounded)							
	1980	1987	1988	1989	1990	1991	1992	1993	1994
Total GDP	2,708	4,692	5,050	5,439	5,744	5,917	6,244	6,550	6,931
Sector									
Electronics manufacturing	55	83	88	97	95	98	99	112	130
Communications	69	125	132	136	147	154	161	173	188
Motion pictures	6	15	16	20	20	20	20	22	25
Amusements and recreation	14	27	30	34	39	42	48	49	52
Printing and publishing	33	62	69	72	74	76	79	82	86
Sector total	177	312	335	359	375	390	407	438	481
Percentage of GDP	6.53	6.65	6.63	6.6	6.53	6.6	6.52	6.69	6.94

SOURCE: Bureau of the Census, 1996, Tables 685 and 686 (*http://www.bea.doc.gov/bea/gpxind-d.htm++curr*).

1977). Sectors selected include electronics manufacturing, manufacturing, durable goods, electric and electronic equipment; communications (including telephone and telegraph and radio and television broadcasting); motion pictures; amusement and recreation services; and printing and publishing. Of these, electronics manufacturing and communications are the dominant categories, with publishing third; so, if dollars of activity represented a viable approach, one would expect to see, in line with the data on information-sector activity presented above, a significant increase in the percentage of GDP represented by these activities. This does not happen; over a 15-year period the increase is from about 6.5 percent to about 6.9 percent. Either the increase in information capacity in the U.S. economy is illusory or these data do not capture the trend.

The latter appears most likely. Some significant sources of error might include the rapidly dropping cost trends in information technology, which, because capability per dollar rapidly increases, makes dollars a poor, and significantly underestimating, proxy for measuring the underlying technology. In addition, much of the information products' value-added comes not from their information content (which is what one wants to measure), but is marked up through the supply chain. Thus, for example, it costs about $1.50 to manufacture a CD, which sells for about $15.00; the $13.50 difference represents packaging, transportation, overhead on stores, profit for various firms involved in distribution, and so forth. Counting this amount for purposes of information density is inappropriate: What is being measured is value-chain economics and profit margins rather than information content. Moreover, a contrary effect also exists: As the automobile example illustrates, much information content is embedded in artifacts and infrastructure systems the economic value of which is captured in these data in noninformation sectors such as manufacturing and transportation. As is the case with any fundamental enabling technology, the use of which is diffused throughout the economy, measures based on sectoral data are unlikely to be sufficiently correlated to be useful.

Accordingly, a less ambiguous alternative measure would be attractive. An interesting possibility is simply tracking the number of appropriate professionals, such as software engineers, available in the economy, on the grounds that, whatever the sector, most information systems require at some point the development of software as a critical input. Unfortunately, this approach also appears to be flawed because available data appear to be inadequate, and for more fundamental reasons as well.

An obvious way to attempt to capture this effect, for example, is to review the educational output. Here the most appropriate categories by which data are collected are communications, which includes associated technologies, and computer and information sciences. The results, shown in Table 6, are ambiguous: The number of bachelor's degrees earned in computer and information sciences in the United States, for example, actually has fallen significantly since 1985, although some of this effect may be explained by the increase in the number of

TABLE 6 Degrees Earned in Information-Related Fields, 1971–1993

Degree	1971	1980	1985	1990	1993
Bachelor					
Communications					
and technologies	10,802	28,616	42,002	51,308	54,706
Computer and					
information sciences	2,388	11,154	38,878	27,257	24,200
Masters					
Communications					
and technologies	1,856	3,082	3,669	4,362	5,209
Computer and					
information sciences	1,588	3,647	7,101	9,677	10,163

SOURCE: Bureau of the Census, 1996, Table 302.

master's degrees in the same category. Nonetheless, overall growth in educational output in both fields remains surprisingly small, probably indicating the fact that capability in both fields can be developed from a number of educational backgrounds.

Nor are employment data necessarily any help. Table 7 shows that, on a sectoral basis, employment has shifted somewhat unpredictably, presumably reflecting the extensive industrial reorganizations, changes in working conditions, and productivity increases that have characterized the recent past in the United States. On the other hand, data on employment by occupation show a more robust growth, especially in mathematicians and computer scientists (Table 8), although given the time span, the growth rates are not, perhaps, unusual. The Bureau of the Census (1996) also projects that related employment categories will grow rapidly: The third fastest growing occupation is listed as systems analysts (from

TABLE 7 Employment by Selected Information Industry Sectors

	Number Employed (in thousands)		
Sector	1983	1994	2005 (projected)
Electronics and other electrical equipment	1,704	1,571	1,408
Communications equipment	279	244	210
Electronics components	563	544	553

SOURCE: Bureau of the Census, 1996, Table 642.

TABLE 8 Employment by Occupation, 1983 versus 1995

| | Number Employed | |
Occupation	1983	1995
Electrical and electronic engineers	450,000	611,000
Mathematicians and computer scientists	463,000	1,195,000

SOURCE: Bureau of the Census, 1996, Table 637.

483,999 in 1994 to 928,000 in 2005, medium projection) and the fourth fastest is computer engineers (195,000 in 1994 to 372,000 in 2005, medium projection).

This brief exercise leads to the conclusion that employment data, although informative, are an inappropriate metric for the information density of the economy. Significant changes in productivity, business structure, and technology are not captured by such measures. For example, data compression and more efficient spectrum utilization technologies have significantly boosted the information capacity available from a given bandwidth; thus, one could have the same number of people working with the same asset base and yet have discontinuous increases in the available information transmission capacity per employee. An employee of a chip manufacturer might be producing slightly more chips per hour, leading to a slight increase in measured productivity, but the chip itself will be far more capable, and far faster, than the previous generation. In such a case, the output per employee measured by chip increases slightly, and the output measured by unit of computation power rises far more dramatically. Indeed, such problems bedevil current efforts to improve traditional economic data collection to provide valid insights and information as the economies of developed countries shift from a manufacturing base to a services base (National Research Council, 1994).

In addition to these measurement problems, another trend in information generation casts serious doubt on any employment-based measure of the information density of the economy. That is, of course, the greatly increased capability of tools, which allow virtually any competent lay person to create information, such as home pages on the World Wide Web or e-mail in bulk, or even, for that matter, massive amounts of personal video; information production increasingly is not limited to specialists. Moreover, the clock time of such information is much shorter than with traditional forms: e-mail is routinely deleted out by the recipient, leaving no trace, whereas letters written in the eighteenth century are still used today by historians. Thus, capturing the existence of such information is more difficult; moreover, the choice of a time dimension in considering information generation and flow in the information economy is critical.

Thus, it appears that only a more rigorous approach, focused on actual information content and manipulation within the economy itself, offers the opportunity to develop viable metrics and indicators for the information density of the economy. The difficulties of accomplishing such a task are legion: Consider the infrastructure that is required to capture economic data on industrial activity. Although developing the appropriate methodology is obviously a task far beyond a single paper, an outline for doing so can be suggested.

The first step is to rigorously define the concept of information density and establish appropriate boundaries. Should, for example, the attempt be made to define the information density of the global economy, of developed country economies [e.g., the Organization for Economic Cooperation and Development (OECD) countries], or of national economies? Parallel to this definitional effort, a taxonomy of data requirements should be developed: What data, if they could be gathered, would be necessary to quantify the defined concept? Where developing the data is either technologically or practically unachievable, the boundaries might have to be adjusted (e.g., if reliable data exist only for the OECD countries, one might begin with them despite the global nature of the information industry, perhaps extending the analysis to the global economy through heuristic rules of thumb).

One possible taxonomy, for example, might involve the collection of the following three components of the information industry:

1. storage capacity, including, for example, memory chips, hard drives, CDs, tapes, and books
2. transmission capacity, measured by available bandwidth at all scales, from international networks, to national and local networks, to firm- and facility-specific networks, excepting the chip level
3. information manipulation capacity, including both software and hardware (perhaps measured by number of transistors per chip)

Other taxonomies might be available as well, and some categories might need to have surrogate measures developed (e.g., measuring available bandwidth might be problematic, so the number of available ports in switches in networks might be a surrogate measure). Undoubtedly, developing the appropriate measures will be difficult, but given the fundamental importance of the trend being measured—the Information Revolution itself—it is clearly both appropriate and responsible to begin the process. From the perspective of the industrial ecologist, such a task is a prerequisite to substantive understanding of the phenomenon suggested at the beginning of this paper, the substitution of information and intellectual capital for other inputs into the economy. Without such tools, anecdotes, analogies, and data bites are enticing and alluring but ultimately subjective and unscientific. Here, as in many places in the field, the challenge to the industrial ecologist is to move beyond that stage.

CONCLUSION

There is an intuitively appealing synergy between the Information Revolution and the concept of sustainability, with some basis in theory and some support from trend data. Such a synergy offers the possibility that future economic activity could support both a high quality of life and a desirable, sustainable world. Nonetheless, there are considerable difficulties in establishing a rigorous, rather than anecdotal, basis for this hypothesis, and much work remains to be done before such a hypothesis can be considered robust, much less demonstrated.

NOTES

[1]Automobile data in this paper refer to sedans with four-, six-, and eight-cylinder engines. They do not include sport-utility vehicles.

[2]Notable contributions have been made by industry cooperatives, such as the Industry Cooperative for Ozone Layer Protection, which led in efforts to develop alternatives to manufacturing processes using chlorofluorocarbons, and the Microelectronics and Computer Technology Corporation (MCC), which led an industry study of life-cycle environmental impacts of workstations (MCC, 1993). The first industry primer on DFE was published by the American Electronics Association in 1992, with contributions from experts from a number of firms. The sector's principal professional group, the Institute of Electrical and Electronics Engineers (IEEE), Committee on the Environment, has sponsored an International Symposium on Electronics and the Environment annually since 1993, with a strong focus on DFE and associated management systems (IEEE, 1993–1997). Individual companies also have contributed significantly. For example, the first textbook on industrial ecology and DFE in the world, by Graedel and Allenby, was sponsored by AT&T, and the company devoted a full volume of the *AT&T Technical Journal* (1995) to the subject.

REFERENCES

Allenby, B.R. 1997. An industrial ecology research agenda. Pollution Prevention Review 8(1):17–38.

Allenby, B.R. 1998. Industrial Ecology: Policy Framework and Implementation. Upper Saddle River, N.J.: Prentice-Hall.

AT&T Technology and the Environment (Special Issue). 1995. AT&T Technical Journal 74(6).

Benbrook, C.M., with E. Groth III, J.M. Halloran, M.K. Hansen, and S. Marquardt. 1996. Pest Management at the Crossroads. Yonkers, N.Y.: Consumers Union.

Bureau of the Census. 1996. Statistical Abstract of the United States. Washington, D.C.: U.S. Department of Commerce.

Buzbee, B. 1993. Workstation clusters rise and shine. Science 261(5123):852–853.

Chichilnisky, G. 1996. The Knowledge Revolution. Columbia University Discussion Paper Series No. 9697–06. New York: Columbia University.

Cohen, J.E. 1995. How Many People Can the Earth Support? New York: W.W. Norton.

Craig, E. 1997. Personal communication. Sun Microsystems.

de Geus, A. 1997. Presentation on the knowledge economy at AT&T, Basking Ridge, N.J. September 9.

The Economist. 1994a. Manufacturing technology, center section survey. March 5, pp. 1–22.

The Economist. 1994b. New-age transport: Trains, planes and automobiles. January 7, pp. 96–98.

EIA. 1996. International Energy Outlook, 1996 with Projections to 2015. Washington, D.C.: U.S. Department of Energy.

Floren, P. 1997. Intel making PCs roadworthy. International Herald Tribune. September 10, p. 13.

Graedel, T.E., and B.R. Allenby. 1995. Industrial Ecology. Upper Saddle River, N.J.: Prentice-Hall.

Graedel, T.E., and B.R. Allenby. 1997. Industrial Ecology and the Automobile. Upper Saddle River, N.J.: Prentice-Hall.

IEEE (Institute of Electrical and Electronics Engineers). 1993–1997. Proceedings of the International Symposium on Electronics and the Environment. Piscataway, N.J.: IEEE.

IEEE. 1995. White paper on sustainable development and industrial ecology. IEEE, Piscataway, N.J.

IMF (International Monetary Fund). 1997. World Economic Outlook: A Survey. Washington, D.C.: IMF.

Japan Ministry of Posts and Telecommunications. 1995. Communications in Japan. Tokyo: Japan Ministry of Posts and Telecommunications.

Jurgen, R.K. 1995. The electronic motorist. Spectrum 32(3):37–48.

Krebs, S. 1993. Advanced engine management systems: the key to reduced emissions and improved performance. Siemens Review (Fall):14–17.

Kurtzman, J. 1997. Interview with Paul Romer. Focus on the Future (April/June):24–35.

Lando, D. 1996. Presentation at Industrial Ecology Conference. Murray Hill, N.J., May 22, 1996.

Lewis, T. 1997. Cyberview: www.batmobile.car. Scientific American 277(1):38.

MacKenzie, J.J. 1994. The Keys to the Car. Washington, D.C.: World Resources Institute.

Magretta, J. 1997. Growth through global sustainability: an interview with Monsanto's CEO, Robert B. Shapiro. Harvard Business Review 75(1):78–88.

MCC (Microelectronics and Computer Technology Corporation). 1993. Environmental Consciousness: A Strategic Competitiveness Issue for the Electronics and Computer Industry. Austin, Texas: MCC.

MCC. 1994. Electronics Industry Environmental Roadmap. Austin, Texas: MCC.

National Research Council. 1992. Automotive Fuel Economy. Washington, D.C.: National Academy Press.

National Research Council. 1994. Information Technology in the Service Society. Washington, D.C.: National Academy Press.

Pacific Economic Cooperation Council. 1996. Pacific Economic Development Report 1995: Advancing Regional Integration. Singapore: Pacific Economic Cooperation Council.

Rejeski, D. 1997. An incomplete picture. Environmental Forum 14(5):26–34.

Thompson, M. 1996. The thick and thin of car cabling. Spectrum (February):42–45.

Wernick, I.K., R. Herman, S. Govind, and J.H. Ausubel. 1996. Materialization and dematerialization: measures and trends. Daedalus 125(3):171–198.

Wriston, W.B. 1997. Bits, bytes, and diplomacy. Foreign Affairs 76(5):172–182.

Intellectual Property Rights in Data

JULIE E. COHEN and WILLIAM M. MARTIN

In 1994, Michael Zeidenberg purchased a compact disc containing phone listings and related information compiled from more than 3,000 telephone directories. Zeidenberg copied the data to his own Web site, which he planned to use for commercial purposes. The manufacturer of the compact disc, ProCD, sued Zeidenberg and ultimately succeeded in stopping him from selling the data, but not because of any protection afforded by intellectual property law. Both the district court and the court of appeals agreed that ProCD's database did not merit copyright protection. However, the court of appeals found that the "shrinkwrap license" that accompanied the product was valid and bound Zeidenberg not to redistribute the information (*ProCD, Inc. v. Zeidenberg*, 1996). This decision has proved extremely controversial. Many legal commentators have criticized the court for allowing ProCD to use a mass-market, standard-form license to confer upon itself broader protection than federal intellectual property law would allow. Others have praised the court for allowing ProCD to do what was necessary to protect its investment.

The Seventh Circuit's decision in *ProCD v. Zeidenberg* and the controversy that followed it illustrate both halves of a growing problem concerning legal protection for databases. The first half of the problem concerns the difficult position in which database creators find themselves. Current intellectual property paradigms were not designed for an information economy (Reichman and Samuelson, 1997). Unlike holders of more traditional types of intellectual property, database creators enjoy only limited protection under the federal intellectual property laws, and so have turned to contracts to protect their investments. The second half of the problem lies in the proposed solutions. Both standard-form contracts (shrinkwrap licenses) and legislative schemes that would grant property rights to database creators ultimately may undermine the broader purpose of intellectual property law. Given the particular structure of the database market, granting broad rights in information invites database creators to price according to profit rather than according to cost. These solutions to the problem of database protection may actually discourage knowledge sharing and hinder research and development (Samuelson, 1997).

SHORTCOMINGS OF THE PRESENT FRAMEWORK

The current framework for legal protection of databases is ill-suited to promote progress in the field of industrial ecology and allied disciplines. Industrial ecology looks to develop a systemic understanding of industrial processes, with the ultimate goals of optimizing material use and minimizing pollution and waste. This inquiry depends on access to data about how the industrial complex functions. Much of the data now being generated is not readily available. This problem may be attributable in part to the current framework of intellectual property law, which was not designed to protect and encourage the dissemination of compilations of factual information.

Consider the example of the intelligent vehicle highway system (IVHS). As proposed, this system would use sensors in the road and in vehicles, in conjunction with global positioning system satellite signals, to relay information to and from vehicles (Dingle, 1995). Such a system might collect a tremendous amount of valuable data, which could then be made available to various interested entities. For example, state departments of transportation might want to know how many vehicles use each highway so they could make better predictions regarding road repair and resurfacing. Civil engineers might be interested in information that could help improve highway safety and efficiency. The Environmental Protection Agency (EPA) might want information bearing on pollution levels, such as the number and types of vehicles that travel at particular times of the day. Public transit authorities could use the information to design bus schedules that better reflect commuter demand, and toll collection authorities could use the information for electronic debiting of tolls.

The data could also be valuable to nongovernmental entities. Advertising agencies seeking to determine the best locations for billboards might want to know traffic volumes and average speeds. Companies that produce traffic updates might want access to real-time data for radio news reports. Car manufacturers might use the information to design vehicles better suited to the actual driving conditions found on highways, resulting in safer and more fuel-efficient vehicles.

What part of an IVHS might benefit from patent protection? The electronics that actually monitor the vehicles could be patentable if they operate in a new and nonobvious way. Likewise, the structure of the database might be patentable if the data were stored in a technically novel and nonobvious structure. That protection, however, would not extend to the most valuable aspect of the database: the data.

Even if databases could be patented, however, patent law is not responsive to the concerns of database creators. It can take years for a patent application to be processed. In the case of the IVHS, database creators will want protection the very week or day that the data are first gathered. Also, to obtain a patent, the invention must be clearly defined to the Patent and Trademark Office so that competitors and the public know what the claimed invention encompasses. This can be difficult with a database that may undergo constant change. Data gathered on the day after the patent filing date would render the patent application incorrect, or would be part of a new database.

Copyright law is equally ill-suited to protect databases. Copyright law expressly bars protection for ideas, functional principles, and facts; instead, the purpose of copyright is to protect original expression. In 1991, the Supreme Court held that a compilation of facts is copyrightable only if the selection or arrangement "possesses at least some minimal degree of creativity" (*Feist Publications, Inc. v. Rural Telephone Service Co.*, 1991). For the white pages telephone directory at issue in *Feist,* there was no way to organize the listings, other than the obvious alphabetical-by-surname method, and still have a directory that would be usable by customers. Because the court grounded its holding not only in the Copyright Act, but also in the Intellectual Property Clause of the Constitution, *Feist* places significant limits on the ability of database makers to obtain copyright protection. Quite often, the obviousness or predictability of the selection and arrangement of data equates to usefulness, and therefore marketability. In a real sense, the database most easily copyrighted is the one that is least marketable.[1]

What aspects of an IVHS database might be copyrightable? One example might be a report that summarizes statistics about road usage for a particular month. The particular statistical facts cited in the report could not be protected, but the expressive aspects of the author's arrangement of them, as well as his or her particular expression of the conclusions derived from the facts, could be protected. This level of protection is of little value to the creator of the IVHS database. There is no single report, arrangement, or expression of the data that captures the essential value of the database. A marketable digital database must be able to present data in many different and useful arrangements.

Turning to state law, there are two theories—trade secrecy and the tort of misappropriation—that under some circumstances provide protection for databases. Unfortunately, neither one is well tailored to the needs of database creators.

A trade secret can be any information that provides an economic advantage to a business relative to its competitors. The information cannot be generally known or easily ascertainable, and reasonable precautions must be taken to maintain its secrecy. The formula used to make Coca-Cola is one such trade secret. Courts will sometimes evaluate a company's precautions to determine whether or not the company itself believes the information has value—if the company does not consider it worth protecting, why should the law? In the case of Coca-Cola, the formula is in a bank vault that can be opened only upon instructions from the company's board of directors.

The secrecy requirement is difficult to meet for databases that are designed to be marketed or shared, as in the case of the IVHS. For example, if the EPA acquires emissions information from the IVHS database creator and publishes it on the World Wide Web so that individual neighborhoods can monitor their exposure levels, the information is no longer secret. The database creator must walk a tightrope between preserving secrecy and making the information usable for its customers. The law requires actions that run directly counter to the motivation that caused the database creator to seek legal protection in the first place.

The database creator could seek to maintain a veil of secrecy by relying on contracts that prohibit each customer from disclosing the information and require the customer to adopt precautions against disclosure. This method can be effective when the number of customers is small and customers have no need to share information among themselves. However, a system such as the IVHS, with a complex web of interested parties, quickly becomes ill-suited to a contracts-based solution because of the high transaction costs involved in making and monitoring each agreement and controlling data exchange between customers. To simplify matters, the database creator might prohibit the lateral exchange of information between customers, but this would decrease the value of the data. In short, it is difficult to reconcile the strictures imposed by trade secrecy law with the goals and potential benefits of an IVHS database.

Trade secrecy law, moreover, does not give a monopoly. It simply limits the means that a competitor may employ to learn secret information. Improper methods include economic espionage, deception, bribery, and breach of contract. Once information has been made public by any of these means, however, the secret is lost. The trade secret owner may be entitled to a monetary remedy, but that may be scant comfort if the secret was a source of continuing value. In addition, it is completely legal to use information acquired by proper methods, such as reverse engineering.

The second area of state law that bears on the protection of databases is the tort of misappropriation, which is based on the Supreme Court's decision in *International News Service v. Associated Press* (1918). The International News Service (INS) had been barred from the European theater of combat and so was limited to gathering news of World War I from its competitors. INS operatives copied Associated Press (AP) newswires from bulletin boards maintained by AP affiliates on the East Coast and then telegraphed the stories to INS newspapers on the West Coast. For obvious reasons, AP sought to prevent INS from continuing this practice. The Supreme Court agreed with INS that the news was not subject to copyright, but it nonetheless held that INS's conduct constituted actionable unfair competition because it undercut AP's incentive to gather the news.

For the most part, the rule announced in *INS* has been applied narrowly to protect only "hot news" or other time-sensitive information. Much of the information that is marketed in database form lacks this quality. The IVHS database, for example, would be valuable in part because it would accumulate data over an extended period, permitting longitudinal study of traffic patterns and emissions problems. However, broader application of the misappropriation tort to data that is not time sensitive may be preempted by the federal Copyright Act (*National Basketball Ass'n v. Motorola, Inc.*, 1997).

PROPOSED STATUTORY PROTECTION FOR DATABASES

Because neither federal nor state intellectual property law provides satisfactory protection for databases, lawmakers and database creators have sought to create a *sui generis* statutory regime of legal protection for compiled information. Some of the proposed regimes, however, threaten to be worse than the situation that they are intended to remedy.

Part of the impetus for *sui generis* database protection comes from Europe. In 1996, the European Commission adopted the Directive on the Legal Protection of Databases, which requires member states to enact legislation granting database creators the "right to prevent extraction and/or reutilization of the whole or of a substantial part, evaluated qualitatively and/or quantitatively, of the contents" of a database (EC, 1996). To gain this protection, the database creator must establish only that there has been "a substantial investment in either the obtaining, verification, or presentation of the contents." The term of protection is 15 years, but is renewable whenever the database holder makes "[a]ny substantial change, evaluated qualitatively or quantitatively, to the contents of the database." This proviso makes the term effectively perpetual because a compiler need only add more data in order to renew protection for the entire database.

Noteworthy for the United States is that the Database Directive includes a strong reciprocity provision. Protection afforded by European Union member states under the new legislation will not be available to foreign companies from nations that have not provided comparable protection. American database companies and their lobbying organization, the Software & Information Industry Association, have invoked the European reciprocity provision to justify the enactment of legislation granting broad property rights in compilations of data. Thus far, however, their efforts have been unsuccessful.

Legislation to create a property right in databases was first introduced in Congress in 1996 (H.R. 3531, Database Investment and Intellectual Property Antipiracy Act). This bill would have granted rights substantially similar to those afforded under the European Database Directive. H.R. 3531 differed from the Database Directive, however, in that the Database Directive authorizes member states to enact limited "fair use" exceptions to database creators' exclusive rights, whereas H.R. 3531 contained no such provision. Because of its breadth and inflexibility, H.R. 3531 quickly encountered strong opposition. In particular, organizations such as the National Education Association, the American Library Association, the National Academy of Sciences, and the National Academy of Engineering expressed concern that the bill would undermine the nation's research capability because of the potential restrictions on access to data (Samuelson, 1997). As a result, H.R. 3531 remained tabled in subcommittee for the remainder of the 104th Congress.

Also in the fall of 1996, the Clinton administration and the European Union submitted proposals for a database protection treaty to the World Intellectual Property Organization (WIPO) for consideration at WIPO's December 1996 conference (WIPO, 1996). The treaty language proposed by the United States was nearly identical to that of H.R. 3531. The U.S. Patent and Trademark Office Commissioner Bruce Lehman, who headed the U.S. delegation to WIPO, admitted that the administration's treaty proposals—the database proposal and a proposed copyright treaty, the terms of which also had failed to secure congressional approval—represented "a second bite at the apple" (Samuelson, 1997). Predictably, the proposed database treaty encountered severe criticism from the same organizations that opposed H.R. 3531, as well as from their international counterparts and a number of developing nations. Lacking consensus on what, if anything, to do about legal protection for databases, the WIPO delegates set the issue aside for further study (Samuelson, 1997).

In the 105th Congress, proponents of strong database protection introduced H.R. 2652, the Collections of Information Antipiracy Act. In addition to arguing that the property right contemplated by H.R. 3531 was overbroad, opponents of H.R. 3531 also had argued that the bill would contravene the Intellectual Property Clause of the Constitution, which (per *Feist*) precludes grants of exclusive rights in facts. H.R. 2652 was billed as a response to both criticisms. Ostensibly, H.R. 2652 would have created a misappropriation tort based on specified unfair conduct rather than an absolute property right in compiled data. As written, however, the bill was as broad as the previous one.

H.R. 2652 would have protected any "collection of information gathered, organized, or maintained . . . through the investment of substantial monetary or other resources" against conduct that threatened an actual or potential market for the database. The bill set no limit on the type of data eligible for protection, few limits on the kinds of uses that might trigger liability, and no term after which the protection would expire. The bill did include a fair use exception allowing extraction of data for educational or research use, but the proviso that the use "not harm the actual or potential market" for the database indicated a very limited range of permitted uses.[2] Thus, as a practical matter, the bill was no different from H.R. 3531. It would effectively have granted a monopoly right; a database maker could prevent anyone from extracting, using, or reusing any part of the database deemed "substantial."

H.R. 2652 died at the close of the 105th Congress, but was reincarnated as H.R. 354 shortly after the 106th Congress convened in early 1999. This time, the bill faced competition; the powerful House Commerce Committee backed an alternative database protection bill, H.R. 1858. H.R. 1858 would have prohibited only the distribution of a duplicate of a database in competition with the maker of the original database, and thus would have granted only limited rights to control derivative markets and value-added uses. In addition, it was drafted to preserve substantially greater scope for fair academic and research use of duplicated information. The major national scientific and research associations, including the National Academies, the Association of American Universities, and the American Library Association, also supported H.R. 1858. The database industries and the House Judiciary Committee, however, remained committed to the basic framework set forth in H.R. 354, and the 106th Congress ended as it began, with no resolution of the database protection issue. The chair of the House Judiciary Committee's Subcommittee on Courts, the Internet, and Intellectual Property for the 107th Congress, Rep. Howard Coble (R-NC), has vowed to reintroduce the Collections of Information Antipiracy Act yet again, and to continue seeking strong, property-like protection for databases.

The framework set forth in the Collections of Information Antipiracy Act—a very strong legal monopoly, coupled with a low standard to qualify and a likely infinite period of protection—is problematic for several reasons. First, the sole basis for the proposed grant is substantial monetary investment by the database creator; no showing of innovation is required. This scheme is paradoxical: The protection rivals that afforded by the patent laws, but unlike the patentee, the rights-holder need not demonstrate that the subject matter constitutes a contribution to society. Second, granting broad and perpetual monopoly rights likely will encourage excessive rent seeking by firstcomers. Database creators will be able to deny the use of "their" data in subsequent compilations or applications, conceivably forever. The important social benefits arising from cumulative and sequential innovation will become subject to the profit motive of rights-holders (Reichman and Samuelson, 1997).

A fundamental principle of intellectual property law is that no one should be given a monopoly on facts, ideas, or other building blocks of knowledge, thought, or communication. This principle underlies the idea-expression distinction in copyright law and its corollary, the merger doctrine, which denies protection to expression that is inseparable from the underlying idea. This is also the reason for denying patent protection to basic principles of science, such as Einstein's theory of relativity or the laws of thermodynamics. The Collections of Information Antipiracy Act attempts no comparable separation of protectable and unprotectable aspects of databases, but asks only whether a challenged use is "substantial." Thus, it appears that if the bill ever becomes law, protected databases will contain no substratum of public domain information that would be available to scientists and other researchers without the rights-holders' permission.

THE PRIVACY PROBLEM

An additional legal consideration that bears on the compilation, use, and sale of data is individual privacy. Electronic databases of information pertaining to individual actions and transactions are potentially quite valuable, but also potentially invasive on an unprecedented scale. For example, various third parties might be interested in purchasing personal identifying information from an IVHS database, including retailers of car accessories (to sell such items as mobile phones and sound systems to people with longer commutes), private detective agencies (to track the movements of particular individuals), auto insurance providers (to determine whether those insured are driving safely and within speed limits), and state highway patrols (to catch speeders and car thieves). Each of these uses raises a host of difficult legal issues.

Governmental purchases of information for law enforcement purposes must be assessed according to constitutional standards. For example, the question whether a state may use IVHS data to apprehend speeders depends, at least in part, on whether such action would amount to an unreasonable search under the Fourth Amendment. In addition, some federal statutes limit the kinds of data that the federal government can collect from individuals; it remains to be seen whether these provisions apply equally to government purchases of personal identifying data from third parties. Discussion of these questions is outside the scope of this paper.

In the United States, there have been few restrictions on the acquisition and use of personal identifying information by nongovernmental entities. However, this situation is changing, again because of pressures originating in Europe.

In 1995, the European Commission enacted the Directive on the Processing of Personal Data, which required member states to adopt implementing legislation no later than October 1998 (EC, 1995). The Personal Data Directive provides that every collector or third-party recipient of personal identifying data must be required to disclose its identity and the existence of the data to each individual identified by the data. Individuals must be allowed to access the data, discover the sources and recipients of the data, and correct any inaccuracies. Individuals must also be given the right to opt out of the use or disclosure of personal data for direct marketing purposes, as well as the right to challenge other practices relating to data collection and use. The Personal Data Directive prohibits the transfer of data to countries that lack adequate privacy protection for individuals. When the Directive was enacted, European Union officials indicated that they considered the United States to be one such nation.

In 2000, the European Union and the United States reached agreement on a set of "safe harbor" information practices for United States companies and organizations receiving personal data concerning European Union nationals. It is too early to predict whether the safe harbor policy will be effective, or whether the United States will adopt similar measures to protect the privacy of United States citizens. Thus far, the government has appeared to favor decentralized solutions to domestic privacy problems, such as the adoption of voluntary codes of conduct by database creators, but this may change if the safe harbor effort fails, or if popular outrage at perceived intrusions increases.

DESIGNING APPROPRIATE LEGAL
PROTECTION FOR DATABASES

The European database protection scheme and the protections proposed in Congress have in common flaws that are intrinsic to a property-based view of information. A preferable system would be designed expressly to balance the interests of database creators with those of society, rather than relying on market forces to accomplish this balancing. What might such a system look like and how would it function? J. H. Reichman, professor of law at Duke University, and Pamela Samuelson, professor of law and information management at the University of California at Berkeley, have proposed one such alternative. They call their proposal a "modified liability approach" because it is based on liability rules (Reichman and Samuelson, 1997).

Liability rules differ from property rules primarily in the absence of a right to exclude. For example, a person who has a property right in a bicycle can deny anyone the use of that bicycle. Under a liability rule system, she would have no such right. Instead, she would simply be entitled to compensation for any use of the bicycle by others. The proper amount of compensation could be determined by the bicycle owner, based on her expected costs and desired profits, or by a court in the course of resolving the bicycle owner's claim for damages, or by a government regulatory body.

The modified liability approach proposed by Reichman and Samuelson would consist of two phases of protection. The first phase would consist of a "blocking period" designed to preserve a certain amount of lead time for the database creator. A property rule would apply during this period, and competitors would not be permitted to use or copy the new database without the database creator's consent. Reichman and Samuelson recognize that in traditional manufacturing there exists a period of natural monopoly afforded by the developer's lead time—the period necessary for competitors to duplicate the new product. They conclude that a regime implementing this dynamic in the database industry would permit database creators to recover what may be significant research and development costs. This would prevent the market failure that might otherwise occur if a competitor could appropriate the database, at minimal cost to itself, and then undercut the originator's prices.

The length of the artificial lead time period afforded under the Reichman-Samuelson approach would be very short, however, for two reasons. First, they argue that the market forces that ordinarily would limit a property owner's ability to reap excessive profits do not exist in the information marketplace. Due to high entry costs, there is a tendency in the database industry for market segments to be left unchallenged once one developer has made a substantial investment in that area. As a result, the database industry is characterized by an absence of direct competition. In such a situation, market forces alone cannot be relied on to allocate resources to would-be users according to their fair value. Second, as in any monopoly situation, the database market is threatened by excessive rent seeking. The most direct method for avoiding these market failures is to set a time after which the database owner's right to exclude expires.

The initial blocking period afforded under the modified liability approach would be followed by an automatic license. Absent some other agreement, the database creator would be obligated, at minimum, to share the data with all secondcomers at rates established by a regulatory body composed of industry representatives and government officials. The ground rules for compensation would be designed to promote competition in the database industry and would permit adjustment of the liability framework when necessary because of changed market conditions. Compensation to owners would be tied to two criteria: (1) their costs for initial research and development and ongoing maintenance and (2) an evaluation of the relative significance of the borrowed content and the value added by the secondcomer. If the secondcomer appropriated the entire database and added little or nothing to it, the rate due the original database creator would be high. Conversely, if the secondcomer added substantial value, the rate paid would be low.

Reichman and Samuelson would not establish a strict compulsory license. They envision their framework simply as setting the baseline obligation for each party, while allowing bargaining for different terms. To prevent abuses of market power by database creators, however, they would require binding arbitration if bargaining failed to generate an agreement acceptable to both parties.

CONCLUSION

As we have shown, current legal protection for databases and proposed property-based regimes for database protection are equally unsatisfactory. Current intellectual property law affords insufficient protection for those who invest time, effort, and money in collecting and compiling data. As a result, database creators increasingly rely on broad and arguably abusive standard-form contracts. Proposed solutions based on property rules, however, would vest in database creators extremely broad exclusive rights in basic knowledge, a result contrary to society's interests.

Under a modified liability approach, the database creator would recover its investment in the compilation process, but the data would remain publicly accessible on fair and reasonable terms. By setting reasonable limits on the power of database creators to exclude and/or charge monopoly rents, this framework would serve society's interests in knowledge sharing, research, and development, as well as database creators' legitimate interests in recouping their development costs.

NOTES

[1]Some courts have responded to this dilemma by stretching the definition of original expression to encompass "soft" ideas that are "infused with taste or opinion" about the usefulness of a particular arrangement of data (*CCC Information Services v. Maclean Hunter Market Reports, Inc.,* 1994). Even so, however, copyright remains poorly tailored to this use.

[2]The bill also would have allowed use of compiled data for news reporting purposes and for verification of independently gathered data.

REFERENCES

CCC Information Services v. Maclean Hunter Market Reports, Inc., 44 F.3d 61 (2d Cir. 1994).

Dingle, J. 1995. FHWA, IVHS, and privacy. Santa Clara Computer & High Technology Law Journal 11(1):15–20.

EC (European Commission). 1995. Directive 95/46/EC of the European Parliament and of the Council of 24 October 1995 on the Protection of Individuals with Regard to the Processing of Personal Data and on the Free Movement of Such Data. Geneva: EC.

EC. 1996. Directive 96/9/EC of the European Parliament and of the Council of 11 March 1996 on the Legal Protection of Databases. Geneva: EC.

Feist Publications, Inc. v. Rural Telephone Service Co., 499 U.S. 340 (1991).

International News Service v. Associated Press, 248 U.S. 215 (1918).

National Basketball Ass'n v. Motorola, Inc., 105 F.3d 841 (2d Cir. 1997).

ProCD, Inc. v. Zeidenberg, 86 F.3d 1447 (7th Cir. 1996).

Reichman, J.H., and P. Samuelson. 1997. Intellectual property rights in data? Vanderbilt Law Review 50(1):51–166.

Samuelson, P. 1997. Big media beaten back. Wired 5.03. Online. Available: http://www.wired.com/wired/5.03/netizen.html [1997, March]. Accessed 3/5/2000.

WIPO (World Intellectual Property Organization). 1996. Basic Proposal for the Substantive Provisions of the Treaty on Intellectual Property in Respect of Databases to be Considered by the Diplomatic Conference. Document CRNR/DC/6. Geneva: WIPO.

Information Systems within the Firm

Improving Environmental Knowledge Sharing

DEANNA J. RICHARDS and MICHAEL R. KABJIAN

Evidence suggests that information for environmental management purposes is being collected and stored at a rapid pace.

- The number of commercially available environmental data management programs grew from 200 in 1984 to well over 2,000 in 1996 (Donely, 1997).
- An analysis of several U.S. chemical companies identified as many as 80 distinct software applications and tools per company to manage environmental information (Kabjian, 1996).
- Over 40,000 environmentally related Internet sites are accessible through the EnviroLink Network, "the largest online environmental information resource on the planet" (Knauer, 1997).

Information technology is enabling the capture, storage, and use of data in ways unimagined previously. Tools such as intranets and document management systems enable firms to achieve new levels of information management, collaboration, and knowledge sharing, and facilitate decision-making processes by providing fundamental support for standard work practices.

These trends bolster the argument made by leading management thinkers that the manufacturing, service, and information sectors will be based on knowledge in the future, and that business organizations will evolve into knowledge creators in many ways. Drucker (1993) suggests that one of the most important challenges for every organization in the knowledge society is to build systematic practices for managing a self-transformation.

This paper explores how systematic practices in the use of information technologies are enabling organizations to use knowledge to improve their environmental performance.

CAPTURING ENVIRONMENTAL INFORMATION AND KNOWLEDGE

The availability of a wide range of timely, relevant information plays an important role in environmental decision making. In managing and designing for the environment, information needs to run the gamut from the simple (e.g., emissions data and inventory information) through the more contextual (e.g., best practices and performance metrics), and then to the complex (e.g., life-cycle assessment and supplier-chain management) and the daunting (e.g., societal and equity considerations of sustainable development).

Effective decision making depends on the appropriate data, information, and knowledge being brought to bear on a problem. However, each of these inputs has a different role in supporting the decision-making process. Recognizing the distinctions between data, information, and knowledge—not always an easy task—is crucial to developing management approaches that leverage their relative values. The fictional scenario depicted in Box 1 illustrates these distinctions: Data are obtained by observing and documenting facts; information is obtained by analyzing and processing data; and knowledge requires cognition, experience, and understanding. This simplistic hierarchy is shown in Figure 1.

The examples of environmental data, information, and knowledge shown in Box 2 illustrate some of the difficulties associated with managing information

---

BOX 1
Data, Information, Knowledge, and
Environmental Improvement

Jane Q. Green, an employee of WEBEGREEN, Inc., has been asked to recommend ways to improve the environmental performance of a certain manufacturing process. She starts by collecting emissions and operating data for the process—an important task. She analyzes the information and learns what is being emitted and how efficient the process is. She then talks with the people who directly manage the process to hear their insights. She also reviews descriptions of previous attempts to improve the environmental performance of the process. She contacts colleagues within the company and other professionals she knows who deal with similar processes. She develops innovative solutions. She receives the company's environmental award. Her immediate supervisor now fears that Jane may be in line for his job. Meanwhile, WEBEGREEN's vice president for environment, health, and safety wishes she could clone Jane to replicate her efforts elsewhere in the many other plants the firm operates!

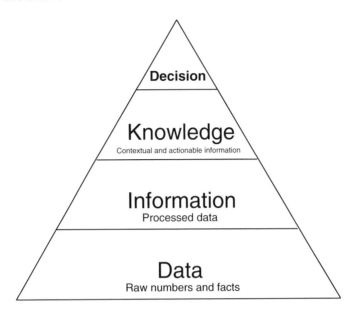

FIGURE 1 Decision hierarchy.

and knowledge. Because information is codifiable (e.g., data and decision or design criteria), information sharing involves the relatively simple process of transmission either through documentation or verbal communication. Knowledge sharing, on the other hand, requires contextual understanding (i.e., it is not

---

BOX 2
Environmental Data, Information, and Knowledge

Examples of Data
- Criteria for air pollutant emissions
- Toxics Release Inventory reports
- Chemical and physical properties of materials

Examples of Information
- Operating and production rates
- Energy and materials efficiencies
- Environmental performance metrics
- Process models showing how planned modifications will affect performance

Examples of Knowledge
- Ideas and strategies to enhance product or packaging composition or design
- Descriptions of past successes and failures in design for environment
- Best-practice guidelines for pollution prevention and waste minimization

codifiable) and is open to interpretation (i.e., it is not articulated easily). Improvements in capturing and managing knowledge in the environmental sphere represent an unexplored opportunity in making improvements in business performance.

Although information management practices in many organizations have a long history and are evolving rapidly, knowledge management practices are somewhat less developed. For example,

- Current knowledge transfer is haphazard in most instances, and there are few tools to support it.
- Knowledge within organizations is scattered, and effective collaboration and knowledge sharing occur inconsistently.
- Institutional memory is short. Very little knowledge is captured and retained for future use. As a result, the same problems are addressed repeatedly by different individuals. In some situations, not being encumbered by history can bring fresh approaches, but in most situations, learning from past experience can be beneficial.

In the case of environmental improvement, vast amounts of information and knowledge have been generated, and many lessons have been learned from successes and failures in addressing environmental concerns. Learning has not been lost—it can be found in best-practice manuals and textbooks, on the Internet, in anecdotes and conference proceedings, and in the memories of people who have worked on the issues. Given the rapid advances in information technology, the key is to more effectively manage and use this information and knowledge.

MANAGING ENVIRONMENTALLY RELATED
INFORMATION AND KNOWLEDGE

The opportunities to improve and apply knowledge management are many, and they cross traditional organizational boundaries. Such opportunities may exist

- throughout a firm's organizational structure (design, manufacturing, environmental, legal, purchasing, accounting, marketing, sales, distribution, customer service, public relations, etc.)
- upstream of operations along the supply chain or life cycle of a product, involving the various suppliers, manufacturers, distributors, customers, and recyclers
- downstream of operations in terms of customer and consumer relations
- in collaborations among the numerous stakeholders, including industry associations, industry peers, government agencies, environmental interest groups, and academia

The objectives and approaches are different in each instance, and certain components of information and knowledge may cut across the four areas described above, but each is relatively distinct within the stakeholder organizations involved (see Figure 2).

Internal Opportunities

Opportunities for effective management of environmental information and knowledge are apparent in a typical firm's operations. In many instances, broad multifunctional teams are called upon to use various knowledge-sharing tools for work related to compliance, product design, production operations, marketing, response to regulatory initiatives, etc. Figure 3 illustrates techniques that may be used to gather and share knowledge at various stages of product development. Figure 4 shows examples of tools used to support information sharing among various manufacturing functions and the types of questions or concerns that may lead to knowledge sharing across work functions.

The tools used to support knowledge management should be designed to meet the varied objectives and diverse backgrounds of team participants who may perform various functions throughout the firm. The tools must be able to capture and translate knowledge derived from projects and other activities and make it available to others within the organization for use in their activities.

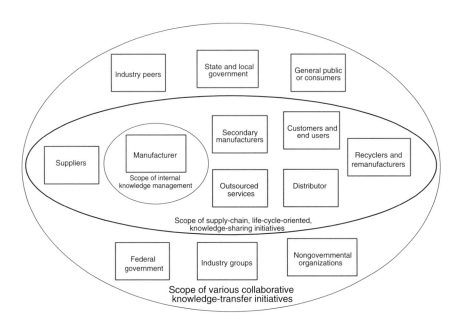

FIGURE 2 Scope of information sharing from the corporate perspective.

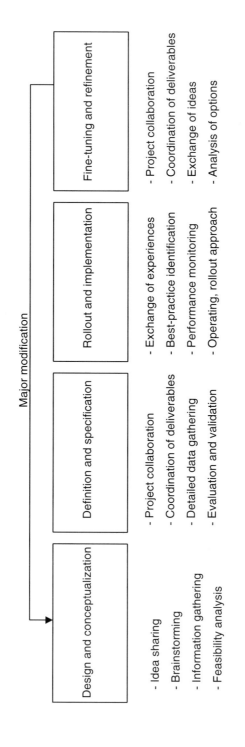

FIGURE 3 Knowledge sharing at various stages of product and project development.

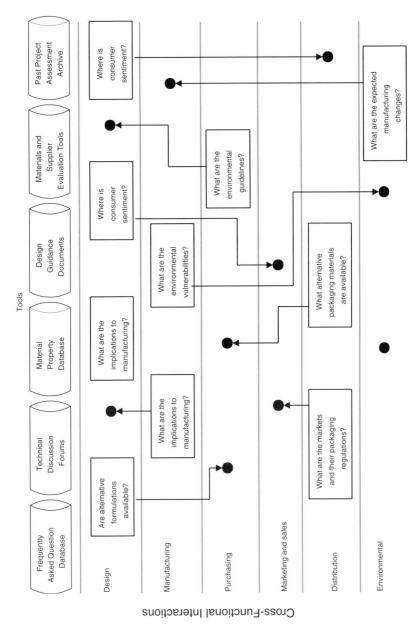

FIGURE 4 Knowledge sharing internal to an organization: product design, manufacturing, and packaging.

Upstream and with the Supply Chain

Today, many firms engaged in the production cycle are more likely to add value to complex production functions by providing services—such as better design, marketing, and distribution capabilities (all information and knowledge value-added activities)—rather than by actually making products. Actual manufacturing is more likely to be accomplished through complex and diverse supplier chains that span the globe. Recent advances in transportation and information technologies have made this model the norm of production functions. To meet production goals, companies have to leverage these techniques, making the complex web of upstream supplier-chain activities operate as one seamless unit. Two other classes of participants are also part of the upstream process: users and customers, who, through their purchasing decisions and patterns, which often are monitored, can help to fine-tune production runs or product requirements. As companies take on responsibility for the environmental impacts of the products they market, the supplier chain that they manage may include recyclers and remanufacturers (who are traditionally thought of as "downstreamers" but who may also supply recycled materials or components).

The wide array of stakeholders upstream in the production function makes the transfer of knowledge daunting. Successful management of this process provides numerous opportunities to identify and exercise options for improving performance, particularly environmental efficiencies. Participants in the process and potential applications of supplier-chain knowledge sharing are shown in Box 3.

BOX 3
Potential Participants and Applications of
Upstream Knowledge Management

Potential Upstream Participants
- Suppliers of materials and equipment
- Manufacturers
- Providers of outsourced manufacturing and related services
- Distributors
- Immediate users
- End users, consumers
- Recyclers, remanufacturers

Potential Upstream Applications
- Supplier evaluation discussions—exchanging evaluations of suppliers and providing direct lines of communication to articulate supplier/customer needs
- Customer information and feedback
- Collaborative development initiatives
- Use of emerging applications of electronic data interchange standards to transfer data more seamlessly between organizations
- Information exchanges that bring suppliers and customers together

At each stage of the product life cycle, stakeholders may exchange knowledge on how to more effectively use, handle, dispose of, or remanufacture a product or material. Effective knowledge transfer along the supply chain can lead to changes in the material composition, in the product design to enable more effective remanufacture, and in the packaging to reduce waste. Figure 5 shows the tools used to support information sharing among the supply-chain players and the types of questions or concerns that may lead to knowledge sharing across functions. Information technology is likely to play an increasingly important role as an enabler of knowledge management; of more effective communication; and of collaboration across organizational lines, borders, and time zones.

Downstream and the Consumers

Downstream factors in complex production operations take on greater significance when services—as distinguished from manufacturing, natural resource industries, and agriculture—are factored into the discussion of knowledge management. Accounting for 60 percent of output and employment (U.S. Department of Commerce, 1996), industries in the service sector provide fundamental economic and societal functions such as transportation, banking and finance, health care, public utilities, retail, wholesale, education, and entertainment. The companies in this sector (e.g., Wal-Mart, Kmart, Target) have great leverage on upstream activities through their merchandise purchasing, in providing food service and delivery (e.g., McDonalds), through their use of logistics and distribution channels to deliver packages [e.g., United Parcel Service (UPS), Federal Express (FedEx)], through the management of hospitals and hotels (e.g., Marriott), in providing health management services (e.g., HMOs), and in providing entertainment (e.g., Busch Gardens, Disney theme parks). Because companies in this sector also interact with a large consumer base, they are a source of knowledge about consumer preferences downstream in the production–consumption system, and they can play a critical role in conveying environmental information to consumers.

The upstream leverage that service firms have on manufacturers is quite evident. As purchasing agents for millions of consumers, these companies exert tremendous leverage over their suppliers by creating markets for environmental improvement. Their downstream influence is yet to be tapped fully. These service firms, to be successful, must be very close to their consumers, and several companies in this sector provide their consumers with environmental information. For example, Starbucks provides information about their environmental practices; Home Depot provides "green" products next to more common brands; and some hotels provide guests with the option to change hotel sheets or towels less frequently to conserve resources. Firms that provide this sort of consumer education also provide early insights into consumer tastes, preferences, and regional buying habits.

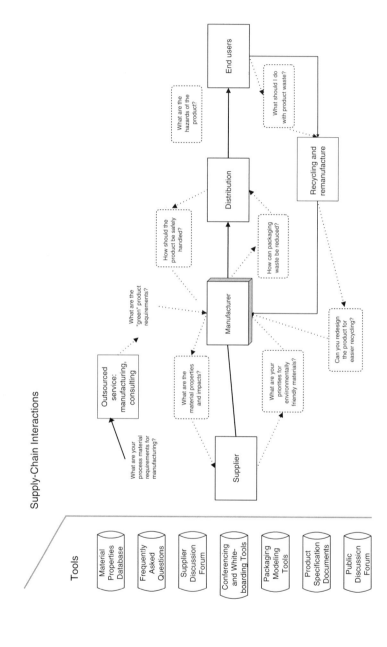

Supply-Chain Interactions

Tools

- Material Properties Database
- Frequently Asked Questions
- Supplier Discussion Forum
- Conferencing and White-boarding Tools
- Packaging Modeling Tools
- Product Specification Documents
- Public Discussion Forum

FIGURE 5 Knowledge sharing along the supply chain.

Many firms in the service sector, and indeed most industrial operations, are also providing environmental education and information about their practices via the World Wide Web. Knowledge sharing (including validation of claims) is an untapped information and knowledge management challenge that involves collaborating beyond the firm with educators, environmental groups, risk communicators, consumer advocates, graphic designers, and information organizers. Box 4 shows potential participants in knowledge sharing in downstream production activities. Figure 6 shows tools used to support information sharing among downstream players and the types of questions or concerns that may lead to knowledge sharing across functions.

Collaborations

Always dependent on information and knowledge sharing, the success or failure of a collaboration depends on a common understanding of the project's objectives and the establishment of trust among collaborators. Information management systems cannot solve these concerns but they can facilitate collaborative efforts for a wide range of objectives. Figure 7 shows examples of questions that may be addressed by an industry group or consortium.

Collaborations are often initiated to address specific concerns:

- **Technology development and diffusion.** The Industry Cooperative for Ozone Layer Protection is one prominent example of an industry

---

BOX 4
Potential Participants and Applications for
Downstream Knowledge Management

Potential Downstream Participants
- General public and consumers
- Nongovernmental organizations (NGOs) and consumer groups
- Government agencies
- Retailers (e.g., Wal-Mart, QVC)
- Food and beverage suppliers (e.g., restaurants, grocery stores)
- Distribution services (e.g., FedEx, UPS)

Potential Downstream Applications
- Consumer information databases containing environmentally related product and service information
- Educational and awareness-building materials (e.g., brochures, courses, projects, games)
- Best-practice knowledge databases and discussion forums—exchange information between downstream service providers, government, and NGOs
- Public discussion forums—exchange information between consumers, government, companies, and others

70

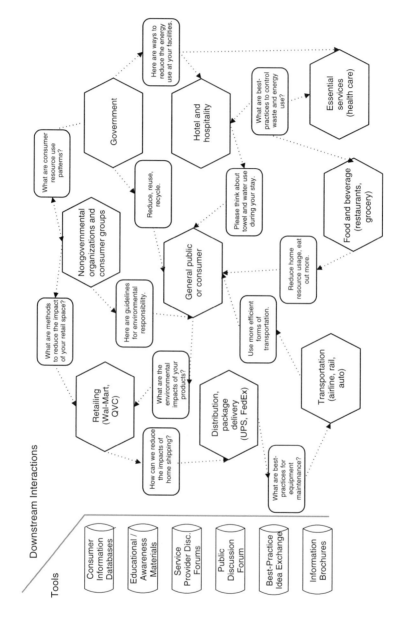

FIGURE 6 Knowledge sharing in the downstream product life cycle (service delivery).

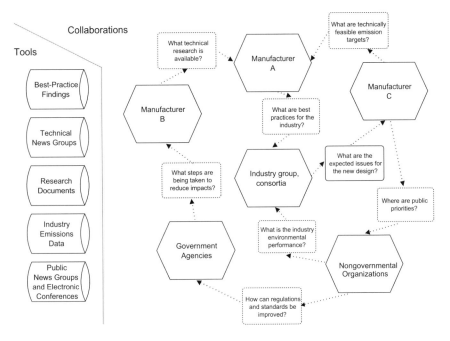

FIGURE 7 Opportunities for collaboration.

collaboration that developed alternatives to the use of ozone-depleting solvents for cleaning electronic components. The leaders of the group, AT&T and Northern Telecom, were industry competitors. Their collaboration brought other companies in the electronics industry together to work on sharing information about alternative technologies. Meant to be a "skunk works" operation, the collaborators worked on technologies that were critical to their firms' operations but that were not key to their market competition. The group also developed an electronic system to share the information they generated. Collaboration among related companies or competitors in the same industry is not rare, and it can fundamentally change the way a business operates. Such collaborations often occur through industry groups and consortia that are intermediary agents in sharing knowledge. Examples of such intermediaries include the Microelectronics and Computer Technology Corporation, Semiconductor Manufacturing Technology, and the United States Council for Automotive Research.

- **Technical assistance.** At the state level, information exchange is critical to enhancing environmental improvement in small and medium-size companies. This information is often garnered with the assistance of universities, environmental groups, and others.

- **Partnerships for innovations in environmental management and design.** These partnerships are modeled on the experiment between McDonald's and the Environmental Defense Fund which resulted in McDonald's re-engineering its processes to substitute paper wrapping for clamshell packing of its food.
- **Regional economic development.** Regional planning efforts often blend economic and environmental concerns to arrive at a consensus. They also can entail an assessment of regional environmental performance. Collaborations based on these issues involve industry, government, academia, and public-interest groups.
- **Fulfilling research agendas.** Whether the issues are global (e.g., global climate change) or local (e.g., water quality), university researchers, environmental groups, government agencies, and industrial firms are key players in defining the problems and arriving at solutions.

Figure 8 illustrates the links among various stakeholders and the potential collaborations that could address specific functional needs.

HARNESSING INFORMATION MANAGEMENT SYSTEMS TO MANAGE KNOWLEDGE

There are many technologies that support knowledge management. Examples of common information technology tools that are used to manage environmental knowledge include:

- **Data management.** Relational database management systems and their associated data management tools provide the ability to store, extract, and analyze large quantities of data arranged in a tabular format.
- **Document management systems.** These systems are used to store text, images, and other types of electronic documents so that they can be more easily searched and retrieved. This provides the ability to create a common repository of documents that is widely accessible and not limited to one location.
- **Groupware and collaborative applications.** Software applications such as Lotus Notes and Novell GroupWise® can include functionality for e-mail, bulletin boards, group scheduling, conferencing, document management, and workflow management. These technologies can help users to work collaboratively and exchange numerous forms of data, information, and knowledge.
- **Networking, intranets, extranets, and the Internet.** This is a broad set of software and hardware technologies that allows computers to connect and share information. The most widely known applications for communications are e-mail and Web browsers (e.g., Netscape Navigator™).

Stakeholder Group \ Functional Activity	Product, Process, Packaging Design	Operating Efficiency	Technology Development	Best-Practice Development	Regulation, Guideline Development	Increasing Public Awareness	Regional Environmental Performance	Consumer Education
Suppliers		•	•	•				•
Manufacturers	•	•	•	•	•		•	•
Outsourced Services	•	•	•	•				•
Distributors	•			•				•
End Users or Customers	•				•	•	•	•
Recyclers/Remanufacturers	•	•		•		•	•	•
Industry Groups	•	•		•	•	•	•	•
Industry Peers/Competitors				•	•	•	•	•
Nongovernmental Organizations				•	•	•	•	•
Government Organizations			•	•	•	•	•	•
General Public and Consumers						•	•	•

FIGURE 8 Overlapping roles in knowledge sharing.

Intranets often involve the use of Web browsers to transfer information within an organization; extranets extend the reach beyond the organization's walls but still limit access, and the Internet can provide general, worldwide access.

- **Information retrieval.** Data miners, intelligent agents, spiders, gophers, and other tools help to retrieve useful tidbits from ever-growing repositories of information. From Web search engines to complex classification systems, these tools help us identify the most relevant data, information, and knowledge for a range of needs.

These technologies are key enablers, but their successful use depends on their ability to support the framework and culture within an organization. To implement these technologies successfully, it is important to know the roles of different organizational groups, how work is performed, and how information flows between groups. The technologies used to manage environmental knowledge are only as good as the organizational structure that supports the work processes through clear roles and responsibilities. The policies, procedures, and guidelines that identify the goals, expectations, and suggested practices within an organization must be clearly articulated and known; only then can the systems and tools that support the framework and culture be successful.

As Heptinstall (this volume) shows, implementing a successful environmental information system requires an understanding of the following steps:

1. **Setting goals.** Determining the environmental goals of a company is the first step in defining the information that needs to be managed. Information management goals for environmental regulatory compliance are going to be different than those for business practices for sustainable development.
2. **Defining processes.** Defining the processes used to generate, retrieve, use, and share information can help to determine the needed infrastructure.
3. **Installing the infrastructure.** Computers and communications technologies will be the largest and most expensive portion of the infrastructure. However, there are also noninformation technology aspects to be considered, such as human networks and knowledge-transfer processes.
4. **Motivating and providing rewards.** Attempts to improve information sharing are doomed to failure unless people are encouraged to share. Policies and cultural environments that reward and encourage information hoarding should be revisited and replaced with compensation contingent on knowledge-sharing activities.
5. **Measuring the results.** Measuring results is a difficult task. It is important to attach milestones and feedback mechanisms to information management projects and to document anecdotal evidence that goals are being met.

PITFALLS IN FACILITATING INFORMATION
AND KNOWLEDGE MANAGEMENT

It is not unusual to have even the best-laid plans run into difficulty along the way. In the case of implementing an effective information and knowledge management system, difficulties may arise in relation to the following factors:

- **Work process changes.** Tools to support knowledge management can require substantial changes to the way a person behaves and performs work. Whether it necessitates communication via e-mail rather than by phone, or referring to electronic documents rather than printed ones, work process changes can be a substantial barrier to success. Tools need to be developed that are compatible with existing, effective work processes. Proactive management of change will be required to improve inefficient work processes and to overcome the traditional resistance to new technologies.
- **Measuring the effects of knowledge management.** The inherent value of knowledge management is difficult to quantify or demonstrate and therefore is often ignored. The most substantial portion of an organization's knowledge assets, or intellectual capital, is embedded in its people as represented by their skills, experience, and intellect. Although the valuation of more traditional corporate assets such as equipment and infrastructure is well developed, it is much more difficult to quantify the value of knowledge assets, and accordingly, to quantify the benefits of preserving them.
- **Implementation.** When implementing a new information system, it is important to focus on how the system supports the use of knowledge in decision making and to avoid the trap of simply collecting knowledge for knowledge's sake. In building knowledge systems, simply collecting information is insufficient; one has to be smart about how to apply the information as well.
- **Hype and distrust.** As is the case in many emerging areas in information technology, there may be more talk than actual practice. As a result, expectations may become distorted and the realized value reduced. Also, fear and distrust are common reactions to the introduction of new technologies. In a survey of management consulting firms—self-proclaimed experts in managing knowledge capital—"less than a quarter of firms in [the] survey said they used the much touted Internet to support" basic knowledge management activities (Reimus, 1997).
- **Hoarding of knowledge.** The notion of knowledge as power can run deep, and the practice of hoarding information is well established in many organizations. To encourage the notion of sharing knowledge, firms need to institute appropriate incentives and reward schemes.

- **Bureaucracy of knowledge management.** Overly bureaucratic approaches to knowledge management can stifle the creative and collaborative processes. To be successful, knowledge management systems must facilitate work processes and be free of unnecessary encumbrances.
- **Overly technical solutions.** A tendency to design approaches that overemphasize information technology will not satisfy real needs. Advanced technology is of little value if it is difficult to use.
- **Legal issues.** Intellectual property and data ownership issues that protect electronic forms of information are examples of legal issues that can hamper knowledge transfer. Also, antitrust concerns can stifle discussions among similar industries working on common solutions. Awareness of these and other issues is important to any knowledge management scheme.

THE IMPORTANCE OF ENVIRONMENTAL KNOWLEDGE MANAGEMENT

The more successful organizations at the turn of the century can be characterized as those that use their soft resources—intellectual capital and knowledge—as effectively, if not more so, than they do their hard assets and infrastructure. Indeed, this trend is true in the economy as a whole, where 60 percent of employment and output is in the services sector (U.S. Department of Commerce, 1996).

In services as diverse as consulting, retailing, air transportation, hotel management, real estate management, freight transportation, and entertainment, knowledge is a prime commodity. Inputs to production are no longer limited to labor, materials, or capital. Technology and information are equally critical. In addition, the systems used to manage complex enterprises are also information intensive. Finally, the values that are being factored into business decisions are no longer as simple as the notion of profit, but include less tangible factors such as the environment, knowledge, and sustainable development.

In addition, there are a number of trends that point to the potential drain of knowledge and the deterioration of organizational memory within corporations. These include

- **Increased corporate outsourcing and contracting.** Although outsourcing and contracting can bring in new knowledge from outside corporate boundaries, they can also deplete the internal knowledge base. Design, manufacturing, and environmental services are prime areas where the use of outsourcing can affect environmental efficiencies. This creates a need to enhance knowledge sharing and collaboration across organizational boundaries and with those providing outsourcing and contracting services.
- **Rapid employee turnover, downsizing, and early retirement.** These business trends heavily affect both the environmental and the information technology fields. Individual employees can possess vital and almost

exclusive knowledge of certain activities within an organization. It can be difficult, if not impossible, to capture the intrinsic value of an employee's experience and knowledge of the company's successes and failures. In a survey of 80 organizations conducted by the Dutch Management Network, 80 percent of respondents reported that only one or two persons within their organization had knowledge of critical business processes (Brooking, 1996). In the oil industry it is estimated that downsizing and restructuring have resulted in the loss of over 450,000 person-years of experience in the recent past (Elliot, 1997).

- **Dispersement of the work force through telecommuting and globalization.** To accommodate telecommuters and project teams made up of individuals dispersed around the globe, knowledge-sharing techniques must evolve beyond communication that occurs around the water cooler or in a conference room. Barriers caused by distance, time zones, or telecommuting create the need for more effective means of sharing knowledge. Much knowledge sharing is now achieved via electronic communication, which both facilitates and requires more effective knowledge management.

- **Increasing reliance on multifunctional and multiorganizational teams.** These teams, or "communities of practice," break the barriers of departments, companies, and even industries and are focused on functions such as design, logistics, or planning. Because the project efforts are dispersed, these teams require mechanisms that allow the group members to work collaboratively, share information and knowledge, and keep a knowledge base of the group's experiences.

- **Fast-paced work environments.** Even as the need for knowledge management increases, the time needed to build knowledge within an organization is often difficult to muster. It is important to recognize that providing access to relevant knowledge can minimize the loss of time and resources due to covering old ground or reinventing the wheel.

As described above, many of today's changing business trends present challenges to environmental knowledge management. On the other hand, many recent developments provide opportunities for improvements.

- **Improved information and communication technology.** The rapid development and deployment of technologies that facilitate collaboration, knowledge capture, and information dissemination provide the technological underpinnings for new knowledge management systems.

- **Increasing experience and sophistication in addressing environmental issues.** As environmental efforts have grown in recent decades, so have the related bodies of knowledge and experience, and thus the opportunities to benefit from knowledge management. The complex nature of many environmental issues will further enhance innovation in the field and encourage professionals to collaborate on the interdependent aspects of their work.

- **The growing importance of organizational learning and knowledge management.** The management of intellectual capital has become a mainstream idea, aided by books such as *The Fifth Discipline* (Senge, 1990) and *Intellectual Capital: The New Wealth of Organizations* (Stewart, 1997). Indeed, many companies have created new positions responsible for knowledge management, such as vice president of learning and organizational development (at Canadian Imperial Bank of Commerce), director of intellectual asset management (at Dow Chemical), and chief knowledge officer (at Ernst and Young).

CONCLUSIONS

The appropriate use of data, information, and knowledge is fundamental to improving environmental efficiencies of production and consumption. Whether information transfer occurs within a company, between customers and suppliers, or among competitive organizations in an industry, the management of information is key to the management of environmental efforts. The rapid growth of information technology continues to provide more effective tools to support knowledge management and transfer. New information tools offer much promise, but in and of themselves are not a panacea. The effective management of environmentally related information is brought about not by making available large quantities of information, but by delivering information that is appropriate for the decision-making tasks at hand. Although many efforts to date have focused on data management, substantial opportunities exist to leverage available knowledge to address environmental performance issues. By undertaking a balanced approach that incorporates data, information, and knowledge, we can begin to more effectively support environmental decision-making objectives as well as longer-term sustainable development goals.

REFERENCES

Brooking, A. 1996. Intellectual Capital: Core Asset for the Third Millennium Enterprise. London: International Thomson Business Press.

Donely, E. 1997. Trends in Environmental Management Software. Paper presented at Environment, Health, and Safety Management Information Systems Conference, Washington, D.C., March 11–12.

Drucker, P.F. 1993. Post-Capitalist Society. Oxford: Butterworth.

Elliot, S. 1997. APQC conference attendees discover the value and enablers of a successful KM program. Center View 5:5–7. Houston, Tex.: American Productivity and Quality Center.

Kabjian, M.R. 1996. Current Approaches and Trends for Environmental Compliance Systems. Paper presented at the American Institute of Chemical Engineers Spring Meeting: Celebrating the Parade of Technology, Washington, D.C., February 25–29.

Knauer, J. 1997. The EnviroLink network. Personal communication.

Reimus, B. 1997. Knowledge Sharing Within Management Consulting Firms. Fitzwilliam, N.H.: Kennedy Publications.

Senge, P.M. 1990. The Fifth Discipline: The Art and Practice of the Learning Organization. New York: Doubleday.

Stewart, T.A. 1997. Intellectual Capital: The New Wealth of Organizations. New York: Doubleday.

U.S. Department of Commerce. 1996. Service Industries and Economic Performance. Washington, D.C: U.S. Department of Commerce.

Using Environmental Knowledge Systems at DuPont

JOHN CARBERRY

The challenges to managing environmental information in a corporation include anticipating the ebb and flow of environmental issues, identifying those issues that are most likely to have a lasting effect, and addressing their restrictive or opportunistic effects on the company. For example, in the late 1980s public opinion of the chemical industry was at an all-time low. There were frequent negative articles in the mainstream press, significant public outrage stimulated by local newspaper reports of environmental incidents, and a high degree of government and academic skepticism. Environmental activists were scaling buildings, occupying sites, disrupting annual meetings, and calling for boycotts.

For DuPont it was a period of awakening to some real problems. The situation was not conducive to conducting business or attracting quality professionals. The company began to recognize that the public's expectations regarding environmental matters were strategic to its success. It was also becoming clear that end-of-the-pipe laws and treatment technologies, prescribed by government in response to public concern and environmental issues, were leading to increased use of capital and engineering personnel for purposes that did not generate revenue. DuPont's cost for compliance was increasing at about 7 percent per year and totaled about $1 billion per year—a staggering amount of money to spend on activities that did not produce products or technology. This amount was even more staggering because it was almost equal to the entire research and development (R&D) budget of the company.

Addressing the challenges of the environmental arena, however, turned out to be easier said than done. The sheer volume of information on environmental issues was daunting at best. DuPont was in a "target-rich" environment, with thousands of issues that potentially could be addressed. To make any progress, it became clear that the company needed to prioritize the most critical issues and then develop a common lexicon that could be used to institutionalize its environmental commitment. Over the course of a year, the company went through a very difficult process that resulted in the development of a set of environmental goals (Box 1) and a corporate commitment statement (Box 2).

Through this process, DuPont was able to translate its environmental priorities into statements that managers at every level could use to make operational decisions. Ultimately, this process resulted in the focus of DuPont's environmental activities shifting from remediation to pollution prevention. Box 3 shows the company's environmental efforts that are focused on pollution prevention. With the development of standardized treatment and remediation technologies along with the shift to preventive strategies, DuPont's compliance has remained excellent, and its environmental costs have decreased. However, with this evolution has come a whole new range of challenges and opportunities for managing information to address the company's environmental priorities.

ENVIRONMENTAL INFORMATION TOOLS

In environmental programs, information overload is an ongoing challenge that DuPont is addressing in a variety of ways. Box 4 shows the technologies that

---

BOX 1
DuPont Environmental Goals

REDUCE
- Environmental and transportation incidents
- Lost workday cases
- Recordable injuries and illnesses
- Airborne carcinogenic emissions
- Listed Toxics Release Inventory "wastes"
- Emissions of 17 large-volume chemicals
- Deep-well disposal
- Packaging waste
- Emissions of greenhouse gases
- Energy use

PROMOTE
- Habitat enhancement
- Production of hydrofluorocarbons and fluorocarbons to replace chlorofluorocarbons
- Double-hulled tankers
- Double-containment fuel systems

SOURCE: DuPont (1997).

BOX 2
The DuPont Commitment

Highest standards of performance, business excellence
- Goal of zero injuries, illnesses, and incidents
- Goal of zero waste and emissions
- Conservation of energy and natural resources, habitat enhancement
- Continuously improving processes, practice, and products
- Open and public discussion, influence on public policy
- Management and employee commitment, accountability

SOURCE: DuPont (1997).

---

are being used to facilitate the exchange of information and to help DuPont's technology organizations integrate the company's environmental priorities into their day-to-day work. Although publications, e-mail networks, teleconferencing, and videoconferencing remain the communication workhorses, external and internal Web-like systems and shared electronic environments are rapidly gaining considerable use.

One example is a system DuPont calls "Having Everything About Remediation Technology," or HEART. HEART is a CD-ROM-based information system that serves as a central source of information on all of the company's remediation technologies. A significant drawback to the information being on CD-ROM, however, is that it cannot be updated readily. The company also has two information databases, on wastewater biotreatment and heavy-metals containment technologies, that are available on internal mainframe computers and easily updated. Both the CD-ROM and the databases are focused on meeting the company's pollution control and remediation needs. Users of both systems are provided with calculation procedures, strength and weakness analyses of various choices, and default recommendations.

BOX 3
DuPont's Pollution Prevention Efforts

- Biosphere impact
- Potential persistent toxins that tend to bioaccumulate
- Alternatives to treatment
- Plant water-use reduction
- Chlorine/fluorine recycle and reduction
- "Green" unit operations
- R&D guidance
- Life-cycle assessment
- Process renewal

BOX 4
Major Information Systems

INTERNAL
- E-mail, teleconferencing, and videoconferencing*
- Technology maps and related hard-cover paper reports for air emission abatement technology
- Programmed mailings of technology summaries to group members and potential users
- An expert system for remediation technology on CD-ROM.
- Expert systems for wastewater biotreatment and heavy-metals containment technologies accessed through an internal electronic network
- Management checklists and approval structures such as PACE and a design checklist for consideration of environmental issues
- Use of internal Web pages*
- Active and passive shared electronic environments*
- Electronic libraries, both central and desktop*

EXTERNAL
- E-mail*
- CD-ROMs for large expert systems, historical databases, or collections of publications*
- Web pages for expert systems or information that is revised frequently*
- File transfer protocol process—"slides" for talks
- Hand-carried diskettes

---

* Emerging winners.

On the pollution prevention front, DuPont is codifying environmental knowledge by incorporating environmental "learning" into its product preapproval procedures. This is done through the Product and Cycle-time Excellence (PACE) system, which integrates environmental considerations into each of the four or five review stages that a new product goes through (referred to as "gates" by Graedel, this volume). At each stage, PACE is used to assess a product's viability on the bases of cost, quality, safety, and potential return on investment. As learning progressed, environmental criteria have been added. Plant designers use a similar checklist when reviewing process designs. The company is also actively looking for ways to extend this approach farther upstream into its R&D efforts.

CHALLENGES IN SHARING REMEDIATION AND POLLUTION PREVENTION INFORMATION

As DuPont's strategic environmental efforts shifted from remediation to pollution prevention, the type of information needed and the nature of information exchange also changed. Information about remediation and treatment technologies was shared easily through broad industry links and developments with other companies. It was nonproprietary and highly specific. Information about pollution prevention, however, is more broadly based and involves (in many instances) the redesign of basic production processes. This information can be highly proprietary, which often conflicts with right-to-know requirements and public access to information.

For example, DuPont makes a product called Tyvek®, a nonwoven fabric material used to make various products, including disposable garments, high-security envelopes, and insulation to wrap buildings. The environmental good news is that Tyvek® is made of high-density polyethylene (HDPE) and contains a significant fraction of postconsumer recycled materials (DuPont used recycled HDPE milk jugs as its raw material for Tyvek®). However, there was once a serious problem with manufacturing Tyvek®, because the original process used chlorofluorocarbon solvents—chemicals banned in developed countries by the Montreal Protocol. DuPont had two options: Get out of the business or find an alternative. Finding an alternative was a monumental task that took some break-through efforts on the R&D front. Indeed, DuPont's leading competitor was unable to solve the problem, and the information about the new process had to be guarded very carefully. Some of the information was soon divulged, however, when DuPont was required to file a revised air permit and submit a new Toxics Release Inventory (TRI) report under the Environmental Protection and Community Right-to-Know Act (EPCRA). This gave the competition key information about the new process. In this particular case, access to TRI-related information was not in DuPont's interest.

However, such access is a two-way street, and in other cases DuPont has used publicly available environmental information about other companies to find potential customers. DuPont's pursuit of zero-emission plants is illustrative. In December 1984, a tragic explosion in Bhopal, India (at a facility not connected with DuPont), released large quantities of methylisocyanate (MIC). The shutdown of existing MIC plants for safety checks meant that every product based on MIC was potentially out of business. MIC shipments ceased, and there was a possibility that MIC permits would not be reissued. DuPont raced to develop a process that would manufacture MIC on demand to be consumed immediately (and as needed) in the next chemical step. The resulting process required no transportation, loading, un-loading, or storage and had a total in-process MIC inventory of about one pound. The new approach came close to zero emissions and zero hazard, and, indeed, the technology concept became a key strategy for DuPont. The company drove lessons learned from this effort into other businesses, where the concept has become a core competency for producing chemicals such as hydrogen cyanide, phosgene, and MIC. The competency that developed as a result of this experience is now considered a business opportunity. DuPont can access public information in the EPCRA listing and TRI reports to identify other companies that are handling hydrogen cyanide, phosgene, and MIC. DuPont can then approach them as potential customers for the new technologies.

MEETING THE INFORMATION CHALLENGE

Despite much progress, there is still a lot of room for improving the collection and dissemination of environmental information, both within and beyond the corporate structure. At DuPont, diverse information systems are being used successfully to address many environmental issues, and there is a strong commitment to continue developing and improving this capacity. However, it is important to note that many of these systems address the needs of specific business units, and their relevance to a wider arena is limited. Indeed, this lack of broad application points to a significant issue in environmental decision making.

In industry as a whole, there is a critical need for the development of basic knowledge about environmental impacts. The systems to deliver information are becoming more sophisticated, but they cannot reach their potential until the information itself is better developed. Companies need useful information in a number of areas, such as methods for product life-cycle assessment, data on persistent and bioaccumulative materials, and distinctions between chemicals that are environmentally desirable and those that are not. Perhaps most importantly, there is a need for prioritization on a national level, detailing the types of information companies must have and the protocols for obtaining it.

Consensus on these and other issues will require collaboration between industry and government, and the information must be scientifically valid, publicly accepted, and easily accessible. After all, the information that is collected will determine what and where action is taken to improve environmental performance.

REFERENCE

E.I. du Pont de Nemours & Company (DuPont). 1997. DuPont Safety, Health, and Environment: 1996 Progress Report. Wilmington, Del.: DuPont.

Environmental Information Management Systems at Rhône-Poulenc

JAMES W. HEPTINSTALL

Rhône-Poulenc Inc. (RPI) is the North American affiliate of Paris-based Rhône-Poulenc S.A., one of the 10 largest chemical companies worldwide. RPI employs approximately 6,000 people and manufactures basic, specialty, and agricultural chemicals at more than 40 locations in the United States and Canada.

In the late 1980s and early 1990s RPI made several large acquisitions, growing from five manufacturing sites to more than 60, and from less than a half billion dollars in sales to over $2 billion. In the mid-1990s the company began restructuring its manufacturing operations and business organizations and reengineering its customer service, procurement, and supply chain processes.

In 1994, RPI decided to reengineer various internal staff functions, including human resources; engineering; communications; and health, safety, and environment (HSE). At that time the company structure consisted of a corporate group and three operating divisions (specialty chemicals, basic chemicals, and agricultural chemicals), each with its own internal functional support staffs. This paper highlights the reengineering of the HSE function and the role that RPI's information management system (IMS) played in the implementation of the redesigned processes.

REENGINEERING THE HEALTH, SAFETY, AND ENVIRONMENT FUNCTION

Phase 1: Discovery

In the discovery phase of the reengineering effort, an RPI project team reviewed the HSE function and defined 15 distinct processes (see Box 1). The team

BOX 1
RPI's 15 HSE Processes

- Management oversight
- Process safety
- Regulation tracking
- Product stewardship
- Training
- Technology transfer
- Permitting
- Environmental services purchasing
- Compliance monitoring
- Reporting
- Medical monitoring
- Workplace safety
- Auditing
- Recordkeeping and retention
- Remediation management

interviewed representatives of RPI's various HSE customers to determine their views of each HSE work process, in terms of its current usefulness and potential future value. In addition, HSE personnel were surveyed to determine the resources (human and material) required to deliver each of the processes. From this information, priorities for the HSE redesign were established and potential costs and benefits were projected.

Phase 2: Process Redesign

A senior management team selected 8 of the 15 HSE processes (see Box 2) for potential redesign. Their selection was based on cost-savings opportunities and the desire to consolidate the HSE function into a shared service organization

BOX 2
HSE Processes Selected for Redesign

- Regulation tracking
- Technology transfer
- Permitting
- Compliance monitoring
- Reporting
- Auditing
- Recordkeeping and retention
- Remediation management

that would serve customers in each operating division. The selected processes represented about 40 percent of the human resources and costs identified by HSE during the discovery phase.

The redesign of the eight selected processes was conducted over six months by a team of HSE professionals and manufacturing managers with the assistance of a consultant. The redesign involved mapping the HSE processes, identifying customer satisfaction attributes, identifying process breakdown points, and recommending improvements. Tools used in the redesign effort included customer surveys and interviews, industry benchmarking, and consultation with subject-matter experts. Specific activities identified as opportunities for improvement included outsourcing; simplifying or eliminating work; and standardizing, consolidating, and automating work. One person from the team was assigned to each HSE process and was responsible for coordinating its redesign efforts.

During the third month of the redesign, RPI reorganized into two operating companies with their functional staff support groups combined into a shared service organization. The HSE function was consolidated from four groups into one, which heightened the need to improve its process efficiencies.

A critical element in seven of the eight targeted process redesigns was the implementation of an IMS that would increase the availability of HSE information throughout the company, improve access to HSE best practices and process information, allow data and documents to be entered once and used by many, improve the speed and consistency of HSE processes, reduce administrative and non-value-added work, and reduce the redundancy of HSE information reporting and recordkeeping.

Phase 3: Pilot Program

A small, four-month pilot study was initiated to evaluate the estimated benefits of the IMS and to determine what type of system was best suited to RPI's HSE needs. Four manufacturing sites agreed to participate in the pilot, and meetings were held with application providers and developers to discuss potential system solutions. It quickly became apparent that customized applications developed for RPI's needs would be more cost-effective than purchasing off-the-shelf applications; the development cost could be divided by 40 (the number of sites), whereas off-the-shelf application costs would be multiplied by 40. An application developer was chosen and six applications (see Box 3) were targeted for the pilot.

Within six weeks the hardware and software were purchased, 80 percent of the applications were developed, and the system was rolled out to the pilot sites. The sites were asked to use and evaluate the applications, and after two months the HSE personnel were brought together again to discuss their findings. Their evaluation involved brainstorming possible areas of savings, quantifying the savings in each area, and verifying the consistency and accuracy of the saving estimates.

BOX 3
Six Pilot Applications

- **Knowledge**
    Distribute corporate and site reference information
    Provide online forum for discussion

- **Incident Reporting**
    Standardize reporting
    Provide information to all sites

- **Site Profile**
    Standardize site information
    Provide information to all sites

- **Material Safety Data Sheet (MSDS) Management**
    Provide desktop PC access with search capability
    Document nonproduct chemicals at sites

- **Document Management**
    Distribute manuals in online (paperless) format
    Disseminate easy updates rapidly

- **Training Management**
    Identify required training by job description
    Track training

The evaluation indicated that the project had a payback of less than a year. The pilot results were presented to senior management, and the project was approved for companywide adoption over two years.

THE HEALTH, SAFETY, AND ENVIRONMENT INFORMATION MANAGEMENT SYSTEM

The HSE IMS consists of a network of 32 site servers. Sites with servers are connected through a local-area network (LAN), and sites without servers are connected either through a wide-area network or via remote dial-up. Currently, more than 1,000 users can access the system using a standard Web browser.

The system consists of national databases (those replicated to all servers), enterprise databases (those replicated to specific enterprise servers), and site databases (those maintained only on site servers). The contents of the national and site databases are listed in Box 4.

Training on the system is provided by an implementation team and is augmented with LAN-based training modules. In addition, each application has its own help section that provides information about the application and instructions on how to use it.

BOX 4
Applications and Databases

National Databases
  Knowledge database
  HSE monthly statistical reporting
  Accident and injury reporting
  Site profile
  Video library
  Notes help
  MSDS catalog
  Database library
  HSE skills inventory
  Training content and delivery
Site Databases
  MSDS management
  Training
  Site documents
  Management of change
  Action tracking
  Inspection planning and tracking
  Process hazard analysis
  Toxics Release Inventory and Chemical Manufacturer's Association
    reporting
  Work-order safety instructions

Following are descriptions of some of the databases in the HSE IMS and how they are being used to share information and knowledge across the company.

Knowledge Database

The Knowledge Database includes a reference section with all of RPI's corporate policies, standards, procedures, model programs, guidance manuals, and interpretations of regulations. Each topic in the reference section has a designated technology manager who is responsible for the content. In the past this information was available only from hard-copy manuals that were updated when enough revisions were made to justify the cost of reprinting and distribution. With the Knowledge Database, revisions and updates are made by the content owner on the site server and replicated across all servers within 24 hours, so employees always have access to the most current information.

The Knowledge Database also has a companion discussion forum where questions can be raised and discussed by interested parties. For example, if someone has a question about Title V compliance, he or she would enter the question in the appropriate discussion forum, where it could be reviewed by personnel

from all sites. Employees with experience in the matter at hand could then participate in the discussion and provide assistance with resolving the issue.

Accident and Incident Reporting

A key objective of the HSE IMS is to share information about accidents and
incidents that occur at various sites in order to help avoid similar occurrences at
other sites. The IMS accident and incident reporting application was developed
for this purpose. It uses a standard reporting form that requests specific information, such as a description of the incident and the results of the initial investigation. The form is replicated to all servers so other site readers can view the
information and then discuss it in small group meetings. This provides a valuable
format to learn how to prevent similar accidents or incidents.

Site Documents and National Site Profile

The HSE IMS also maintains site-specific documents, such as safety manuals, policies, operating instructions, operator logs, and procedures. As with the
corporate-related documents on the Knowledge Database, the IMS allows almost
instant revisions to site-specific documents, thereby eliminating the cost of duplication and distribution.

Other information about sites is available in the national Site Profile Database.
Templates created for this application provide a standardized format for collecting
and displaying site information. Now, instead of having to locate a contact to obtain
information about a site, users can simply access the site profiles on the database.

Material Safety Data Sheets

At the site level, the greatest impact of the IMS has been in the creation of a
Material Safety Data Sheet database. Before automating the collection of MSDS
data, each site maintained a series of hard-copy books (from one to six) for all the
nonproduct chemicals handled at the site. These MSDSs were supplied by the
chemical provider. Most sites had to maintain multiple copies of the MSDS
books because they were required in various locations (control rooms, maintenance shops, labs, etc.) throughout the facility. This system required someone at
each site to maintain the MSDS books (adding, copying, and distributing new or
revised MSDSs to all the locations).

Today, MSDSs are scanned onto the IMS so that they can be viewed electronically from any PC workstation. Each MSDS has a cover page that lists the
manufacturer's name and the chemical abstract number, which enables users to
do keyword searches for chemicals. The cover page also has fields for listing
certain critical information such as personal protective equipment, spill cleanup
procedures, and chemical hazards. If a user wishes to view the actual MSDS, a

click on an icon takes the user to the scanned document. This application allows for quick access to the necessary information in an emergency and eliminates the need to maintain multiple hard-copy books.

Additional Databases

Several additional applications have also been developed. They include:

- the Management of Change application for tracking, approving, and documenting process changes or other critical procedures
- the Action Tracking application for tracking and managing action items and due dates established as a result of regulations, incidents, inspections, or other commitments
- the Site Document Navigator application for obtaining quick access to site information (references, policies, procedures, forms, files, etc.)

In addition to these applications, procedures have been implemented for network administration, security, and lost data recovery.

SUMMARY

The HSE IMS project has been a success because the manufacturing sites found it valuable. The efficiencies defined in the business process redesigns are being achieved through the implementation of applications that were identified and prioritized by the manufacturing site HSE personnel.

This approach, working from the site-level upward, created momentum that carried the project through its pilot phase to implementation. It also allowed for close scrutiny by those responsible for corporate information technology and general management.

The implementation of the IMS by the HSE organization led to cost savings via personnel reduction (due to attrition). The implementation process also brought the HSE function physically closer to its customers (manufacturing sites) in the development of the system, and today maintains a way for HSE to be "virtually" close to its customers in distant locations.

In addition, the IMS infrastructure provided manufacturing sites with information-sharing capabilities well beyond HSE issues. For example, some sites have initiated discussion databases, shift-report databases, and other similar applications. As for the HSE community, the speed of communication that the IMS provides has enhanced the transfer of technology from external resources found on the Internet and from other professional sources.

As a final note, RPI has experienced continued declines in its lost workday and employee injury frequency rates. The IMS has played a small part in this accomplishment by providing rapid distribution of HSE experiences, information, and knowledge.

Environmental Knowledge-Sharing in Manufacturing

THOMAS E. GRAEDEL

Although a wealth of environmentally related information exists within the modern corporation, it tends to be diffusely distributed. For example, one person might know the kinds and amounts of materials that a company purchases and uses. Another person might know the energy consumption of the manufacturing facilities, perhaps down to that of individual manufacturing lines. Other individuals may know about the wastewater treatment processes, the atmospheric emissions, or the process by-products and their values. Still others may know how new products are designed and how readily those products can be recycled. No one person, however, is likely to know all of these things.

Such environmental information is commonly used in corporations for reporting purposes, to verify utility billing, or to ensure that orders are placed for materials used in manufacture. What is less common is to see this information used as an integral part of corporate decision making, although it could and should be. The ways in which this might be achieved are the subject of this paper.

THE GATE CONCEPT IN MANUFACTURING

Modern industrial managers wish to stimulate their design and development staffs to generate numerous ideas for new products, in the hope that a few really successful products will result. However, carrying every product idea through from concept to manufacture is too expensive to be feasible, so a structured process, the "product realization process (PRP)," has been developed to guide business decisions along the way (Ulrich and Eppinger, 1995).

There are a number of versions of the PRP, with names such as "integrated development system" and "integrated product development," and many corporations have developed handbooks to explain them (e.g., Carrier Corporation, 1995; United Technologies Automotive Corporation, 1995). PRP approaches vary in level of detail and in the number of sequence steps, but they all share the general approach, if not each specific step, shown in Figure 1.

Eight steps in the PRP, from idea to obsolescence, are indicated in Figure 1 and described in more detail throughout this paper. The transitions from one step to the next are called "gates," and they are opportunities for managers to decide whether to permit the product development to proceed to the next step. In the formal structure of the PRP, a review is held when a product reaches each gate in the sequence. The review team typically includes representatives from design, manufacturing, purchasing, marketing, and other appropriate corporate departments.

The items considered at each gate review include marketability (Do we still think our customers want this product?); manufacturability (Can we make the product as envisioned?); economics (Can we make a profit on this item?); strategy (Are we ahead of our competitors?); and a variety of other factors. Cost is a major influence on decision making, especially in the later stages of the PRP. As seen in Figure 2, the financial investment required to move to the next step of product development increases as one moves from gate to gate. By gate four or five, if the product is then judged to be unpromising, a substantial unrecoverable investment will have been made. The goal of the review process is to let promising products move quickly to manufacture but to close gates early on projects that will consume investment dollars without the probability of substantial financial return.

PRP gate reviews often omit considerations of environmental issues, largely because tools have not been formalized for bringing such information into the process. Relevant environmental information, therefore, is often not presented even if it is available within the corporation. However, such information can, in principle, be provided at each gate if corporate knowledge sharing is practiced. And, if environmental information is considered in the gate decision, a better overall decision is likely to be made.

ENVIRONMENTAL KNOWLEDGE AT THE GATES

Information of all kinds becomes more detailed as a product progresses from early to later stages of development: Concepts are transformed into designs, materials are specified, sizes and features are determined, costs are calculated more accurately, and customer response can be better estimated. Accordingly, detailed environmental information cannot be provided at early stages, nor is it needed (Hoffman, 1997). As successive gates are passed, however, the environmental information must become more and more comprehensive to be of the most use.

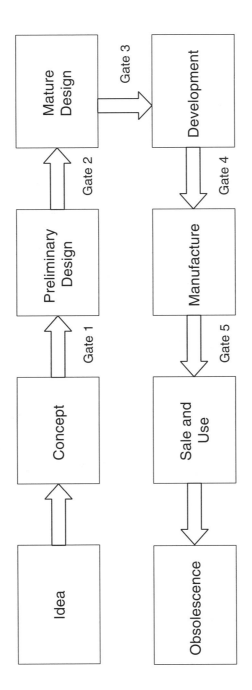

FIGURE 1 The steps in the PRP and the gates that potentially inhibit passage from one step to the next.

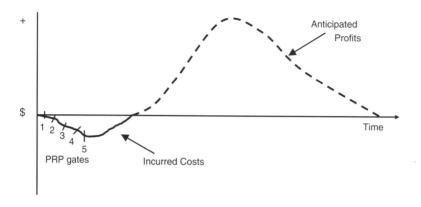

FIGURE 2 A typical schematic diagram of the financial cost and profit performance of a successful product. The locations of the PRP gates are shown along the financial cost curve.

To illustrate these points, Table 1 lists the product and process information available at each gate for a typical manufactured product. The items in this table serve as a guide to the environmental information that can be brought to bear at each gate review.

Gate 1: From Concept to Preliminary Design

The first gate controls the transition from concept to preliminary design. The business questions at this gate are very basic: Does this concept appear to meet a customer need? Is it consistent with the corporate product line? Does it have the potential to compete effectively?

The environmentally related questions are basic as well and are designed to discourage product concepts that involve unfavorable environmental attributes, such as the use of forbidden or highly regulated substances. As shown in Table 1, these questions can be addressed at Gate 1 only for the principal materials and processes.

The appropriate environmental tool at Gate 1 is thus a list of "inviolates": product or process attributes that the corporation has decided will not be permitted. A typical list of inviolates for high-technology product manufacture is given in Box 1. Lists of inviolates include materials, processes, or practices that are illegal or that might involve potential liabilities which a corporation would rather not assume, even if current regulations are not an issue.

Gate 2: From Preliminary Design to Mature Design

The initial or concept stage of product development typically involves a small group of people and the only expense is their time. At the next stage,

TABLE 1 Information Known at Product Development Gates

Product	Process
Gate 1	Key manufacturing processes (with
Principal material(s)	technology and chemicals)
Critical electrical characteristics	
Critical mechanical characteristics	
Size	
Gate 2	
Major components	Principal manufacturing processes (with
Preliminary electrical design	technology and chemicals)
Preliminary mechanical design	
Preliminary visual appearance	
Gate 3	
All components	All manufacturing processes (with
Final electrical characteristics	technology and chemicals)
Final mechanical characteristics	Process energy consumption
Final visual appearance	
Mold designs	
Gate 4	
Final materials list (constituents and	All by-product streams
quantities)	All waste streams
Recyclability	Outside supplier interactions
Packaging	
Gate 5	
Marketing	Shipping

BOX 1
Typical Product and Process Environmental Inviolates

Gate 1 Inviolates
- Chlorofluorocarbons and hydrogenated halocarbons that are restricted by the Montreal Protocol may not be used in any manufacturing process.
- Radioactive substances may not be used in any product.

Gate 2 Inviolates
- Mercury switches may not be used in any product.
- Plastics must not contain additives (colorants, stabilizers, etc.) formulated with the following metals: Ag, As, Ba, Cd, Cr, Hg, Pb, Se.

Gate 3 Inviolates
- Cadmium-plated metal components may not be used in any product.
- Plastic components may not be used without an appropriate International Organization for Standardization symbol.
- Plastics may not contain polybrominated flame retardants.

Gate 4 Inviolates
- Recycled stock must be used for all packaging material and descriptive literature.

Gate 5 Inviolates
- Products may not be advertised as environmentally superior to competing products (but their positive environmental attributes should be pointed out).

preliminary product design, the size of the group expands but the activities are still limited to sketches, conceptual CAD/CAM products, and lists of preferred materials, so the embedded development expense is still modest. At Gate 2 the major design decisions have been made, but few details are available.

The typical business questions at Gate 2 are formulated from the perspective of the preliminary design: Do the estimated performance specifications meet the product goals? Is the design visually attractive? Is the product likely to be profitable? The answers are important because the corporate investment in a product that passes the second gate begins to increase rapidly.

Because the product design has progressed significantly by Gate 2, there is now significant information that can be reviewed from an environmental perspective. The review team can evaluate the environmental aspects of the design approaches for both the product and the process. In some corporations there is little formal guidance for such review, which can make it difficult to evaluate a product's compliance with Gate 2 criteria. However, some corporations have systemized this process. For example, Lucent Technologies (1996) publishes an internal "Designer's Companion," which is a series of case studies of fortunate and unfortunate design choices (some environmental, some not). The result is a manual of design choices that can be reviewed as part of the Gate 2 approval process.

Gate 3: From Mature Design to Development

At Gate 3 the design team presents detailed information on the product design and moderately detailed information on the associated manufacturing processes. At this stage the product can undergo a reasonably thorough environmental review. If the product is relatively similar in type and materials to other products of the corporation, there may be little need for a comprehensive environmental review of the manufacturing processes. If new processes are required, however, the manufacturing review will be more extensive.

The Gate 3 product review is in all cases quite detailed. From a business standpoint the questions become more focused: Are there technical impediments to development? Are the manufacturing processes satisfactory? Are the electrical and mechanical goals for the product fully realized? Will the product have customer appeal?

Environmental information at Gate 3 can be derived from guidelines and checklists for environmentally preferred design decisions (Figure 3 illustrates an excerpt from such a checklist). In many corporations, such tools are now incorporated into the computer-aided design process and has the potential to become part of the designer's product development goals. The Gate 3 review can thus evaluate the degree to which a product design incorporates recommended attributes from the checklists. At this gate there is also enough information available to perform a semiquantitative or "streamlined" life-cycle assessment (SLCA) (Graedel, 1998).

A: Construction and Joining Techniques

Concerned units: control unit, monitor, keyboard

	Requirement	Applies to Assembly(ies)	Cat.	Met? Yes	No
A.1	Are assemblies made of incompatible materials separable or joined by means of separation aids?	Housing, chassis, electronic assemblies	M		

Joints between the housing and chassis and between the chassis and electronic assemblies are important. If the assemblies and materials are to be reused or recycled, these parts need to be disassembled easily. If the components contain toxic substances, separation has to be done quickly and safely. The "compatibility" of materials can be tested, for example, using compatibility matrices.

A.2	Are components containing toxic substances easy to find and simple to remove?	Electronic assemblies	M		

The minimum target for recycling is the removal of toxic substances such as batteries and capacitors. They have to be easily separated.

A.3	Are joints that have to be opened or released easy to find?	Housing, chassis	S		

Connections that have to be opened or released in the dismantling of the product have to be located easily and quickly. If they are concealed, appropriate advisory labels should be affixed to the product.

A.4	Can the product be dismantled using only universal tools?	Housing, chassis, electronic assemblies	M		

The term "universal tools" refers to everyday, commercially available tools.

A.5	Have necessary points of application and working space taken into account the need for space for dismantling tools?	Housing, chassis, electronic assemblies	M		

If tools are needed to engage release mechanisms, adequate working space has to be provided.

FIGURE 3 Sample checklist for recycling-oriented design of personal computers. SOURCE: Adapted from Steinhilper, 1996.

In such an assessment, the entire range of potential environmental impacts is evaluated for each product life stage—premanufacture, manufacture, product delivery, product use, and end of life. By taking advantage of checklists and the SLCA, designers can correct a product's unfavorable environmental attributes before the design is finalized.

Gate 4: From Development to Manufacture

By the time the Gate 4 review committee meets, the design is completed, the manufacturing process is set, the materials and components have been chosen, and the suppliers have been at least tentatively identified. The decision at this gate is whether to proceed with manufacture—the most costly of all the stages.

The business decisions at Gate 4 are obvious and important: Have the cost estimates been met? Is the product manufacturability satisfactory? Has a reliable set of suppliers been identified? Will the final manufactured product retain the desirable characteristics identified at Gate 3?

With the product and process information now finalized, either an enhanced SLCA or a comprehensive life-cycle assessment can be performed (Curran, 1996). Most items of environmental concern will have been identified by Gate 4, but product delivery implications can be addressed in detail for the first time, and the overall results can be made quantitative to the degree desired.

Gate 5: From Manufacture to Sales and Use

The Gate 5 review is often ceremonial, especially if decisions at previous gates have been sufficiently thoughtful and comprehensive. Provided that no unexpected and unwelcome information has arisen, the product is released for sale and use. The business questions involve a review of the degree to which the product manufacturing meets expectations and the ways in which the marketing campaign should move forward.

From an environmental standpoint, questions asked at Gate 5 concern whether environmental issues have been properly reviewed at previous gates, whether the product delivery and marketing activities will meet environmental goals, and whether provisions need to be made for end-of-life activities, such as product takeback or battery recycling. These reviews, and those of earlier stages, are aided by tools such as corporate environmental management protocols or International Organization for Standardization 14000 standards.

CONCLUSIONS

As this paper shows, a wealth of environmental information is available within corporations to aid in the decision-making steps of the PRP. In many cases, however, corporations have not implemented procedures to integrate that information into their decision making. The PRP format provides an important and convenient way to accomplish that integration.

Although PRP gate passage is a discrete and reproducible sequence of actions, the use of the described environmental information tools is less circumscribed. Different corporations and review teams may use variations of these tools or implement them in a different manner. The way in which an individual corporation proceeds will be a function of its environmental management plan. The important factor is not that environmental information is used in a prescribed manner, but that a mechanism is in place to guarantee the use of the information at PRP gate reviews. When that mechanism is established, there is great potential for benefits to the environment and, increasingly, to the responsible corporations themselves.

REFERENCES

Carrier Corporation. 1995. Integrated Development System Project Leader's Guide. New York: Carrier Corporation.

Curran, M.E., ed. 1996. Environmental Life-Cycle Assessment. New York: McGraw-Hill.

Graedel, T.E. 1998. Streamlined Life-Cycle Assessment. Upper Saddle River, N.J.: Prentice-Hall.

Hoffman, W.F. 1997. Recent advances in design for environment at Motorola. Journal of Industrial Ecology 1(1):131–140.

Lucent Technologies, Bell Laboratories. 1996. Designer's Companion. Murray Hill, N.J.: Lucent Technologies.

Steinhilper, R. 1996. Private communication. Fraunhofer-Institut für Produktionstechnik und Automatisierung.

Ulrich, F.T., and S.D. Eppinger. 1995. Product Design and Development. New York: McGraw-Hill.

United Technologies Automotive Corporation. 1995. Product Development and Launch Process Leader's Guide. Detroit, Mich.: United Technologies Automotive Corporation.

Modular Design for Recyclability
Implementation and Knowledge Dissemination

KOSUKE ISHII

"Modular design for recyclability" is a methodology that helps design engineers plan for potential uses of a product after it is retired (Ishii et al., 1992; Marks et al., 1993). The purpose of this method is to make recyclability an integral part of design practice, in step with standard considerations such as manufacturability and serviceability. It is based on the premise that proper consideration of recyclability can enhance the ecology of industry by extending the utilization of resources and by reducing waste, energy requirements, and other environmental loads. This paper explores the recyclability of products based on the author's efforts at the Stanford University Center for Professional Development. That effort demonstrates that modular design provides engineers with a systematic methodology to enhance a product's recyclability, while still addressing more common design requirements.

The successful implementation of modular design depends on several factors. First, it must be integrated with other common design methodologies, such as quality function deployment (QFD) (Hauser and Clausing, 1988) and design for assembly (DFA) (Sturges and Kilani, 1992). Engineers are more likely to use the method if it is linked to their primary objective of realizing functions at feasible manufacturing costs.

Second, data requirements must be met. For example, engineers need data on the efficiencies of various disassembly and sorting methods, on costs to recondition and reprocess components and materials, and on the resale values of reusable components and recycled materials. Of course, this information changes with time, market demand, and geographical location; however, at a minimum, engineers need best estimates of this data for each new product. They also need information about when a product might be recycled and about the logistics involved in recycling it. In particular, knowledge needs to be developed, either within an organization or across an industry, about metrics related to recyclability and how to evaluate products under such metrics.

Third, information about modular design must be disseminated so that it can be integrated into the everyday activities of engineers. At Stanford University this dissemination is done through its distance learning facility. The Stanford Instructional Television Network covers modular design for recyclability in a graduate-level course on design for manufacturability. In this course, students form teams and apply various methodologies to real-life product development examples. On-campus students work on examples provided by various companies, and long-distance corporate students work on projects relevant to their workplaces. An advantage of this long-distance educational effort is that it combines academic research results with best practices from industry, and it delivers this knowledge to a much wider audience than is normally reached by a standard graduate-level course.

Just as it took several decades for quality engineering to be accepted by U.S. industry, it is expected that it will take some time for companies to truly integrate environmental concerns into their activities. However, the integration of environmental considerations into everyday manufacturing activities may be accelerated by linking those considerations to standard product development practices and by disseminating new techniques through mechanisms such as long-distance education.

DESIGNING WITH RECYCLABILITY IN MIND

A recyclability strategy begins by understanding what might occur in a demanufacturing or recycling plant (Figure 1). Clearly, a recycling plant must be able to handle more than one model of a product, so any strategy must be based on the recyclability of entire product lines or families of products. It must also be based on the need to handle multiple generations of products, different model years, and discards from any stage in the supply chain. The strategy must also consider the possibility that products from different manufacturers may have to be processed. Therefore, it is important that designers select materials such that similar materials are used in the entire product family over multiple generations. To enhance component reuse and material recycling, engineers need to embed modular recyclability concepts into product design so that the costs of recycling can be lowered. Successful product recycling can lead to reductions in solid waste and in raw material and energy usage throughout the product life cycle.

Material selection is a critical factor in any design-for-recyclability strategy. Designers can facilitate recycling by using fewer types of materials in their products and by making it easy to separate components that are made from different sets of materials. Designers need to consider these factors across entire

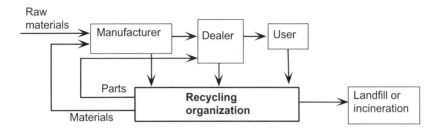

FIGURE 1 Design method focusing on recycling (demanufacturing).

families of products so that one recycling plant can effectively process all of a product's different models and generations.

Although these general design principles have been recognized widely, recyclability is still a low-priority design criterion relative to considerations of function and cost. The challenge, therefore, is to get designers to consider recyclability as part of product design. This challenge is best met when knowledge about recyclability is integrated into standard design tasks and when tools to aid the integration are developed and deployed.

INTEGRATING MODULAR DESIGN FOR RECYCLABILITY INTO PRODUCT DESIGN

To effectively implement recyclability considerations into product design efforts, the process must be streamlined for engineers. Environmental impacts should be reviewed in a standard manner, along with conventional product requirements such as cost and functionality. This is done most effectively when environmental considerations can be integrated into standard design methodologies such as QFD, functional analysis, and DFA.

Quality Function Deployment

QFD is a method that emerged during the 1970s in Japanese engineering efforts. It translates customer requirements into product characteristics, design attributes, manufacturing factors, and process control parameters. It identifies engineering specifications and designs that respond to the specific customer needs. Today, many American companies use QFD as a standard benchmarking and requirements identification tool. One can apply QFD to environmental considerations by identifying the pertinent relationships between environmental requirements and attributes, such as material selection and modularity. For example, environmental considerations in designing a washing machine would include quiet operation, low-energy usage, and minimal weight. Although these items coincide with customer concerns, there are other environmental issues that designers normally do not consider, such as minimizing the use of hazardous materials or precious resources. Many environmental concerns parallel what end users want to see in a product; some, however, are not as obvious. Designers need to force themselves to actively consider environmental impacts as part of the product development process.

Functional Analysis

Functional analysis is a key tool in "value engineering," a practice originally developed in the United States during the 1950s and which contributed significantly to product development in Japan. Functional analysis of a product design results in a hierarchical clarification of how primary and secondary functions decompose into subfunctions and eventually correspond to the physical implementation of the product in terms of its components and subassemblies. Functional analysis is particularly effective for identifying an appropriate modular structure that is based on manufacturability, serviceability, and recyclability. For electromechanical products, in particular, functional analysis helps designers to identify modular parts for products with different technological life cycles, thus promoting reuse and remanufacture of modular parts with longer life cycles.

Design for Assembly

DFA is another popular method that is used by American manufacturers. There are many variations of DFA; however, they all encourage designers to envision a product's assembly steps, identify cost- and time-dependent processes and relate them to design attributes, and seek improvements. DFA is used most effectively at the layout design stage. This is the stage when engineers consolidate assembly structures and select materials but have not yet developed detailed designs for components. At this stage, while analyzing a product's assembly, it is also appropriate to consider the product's retirement, its disassembly, and its processing for reuse and recycling.

An assembly fish-bone diagram is a DFA tool that provides a rough chart of a product's assembly process (Ishii and Lee, 1996). This tool is easily adapted for recyclability purposes as a "reverse" fish-bone diagram, and when combined with an associated recyclability map, can be effective in any process of modular design.

The Reverse Fish-Bone Diagram

The reverse fish-bone diagram is a graphical representation of a product's retirement process (Ishii and Lee, 1996). Figure 2 shows a paper tray for an ink-jet printer and a fish-bone diagram representing the processes associated with recycling it. This example shows a small part of a larger hierarchical fish-bone

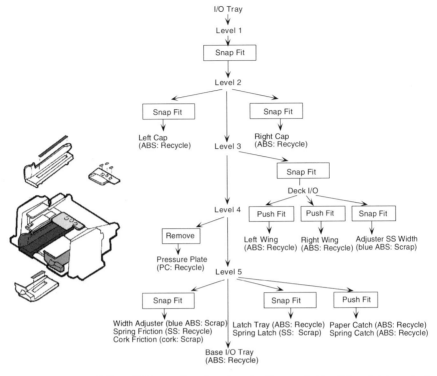

ABS and PC = grades of engineering thermoplastics; SS = stainless steel; I/O = input/output

FIGURE 2 Paper tray and its reverse fish-bone diagram.

diagram for the entire printer. The diagram is the reverse of the assembly fish-bone diagram, which is used to document the assembly process. Fish-bone diagrams can help designers consider both assembly and recyclability criteria at the same time.

The size and shape of a fish-bone diagram indicate the complexity and cost associated with the demanufacturing process. The number of levels in the diagram indicates the number of different disassembly stations required to recycle the part or product. The rectangular nodes (labeled "snap fit," "push fit," and "remove") in Figure 2 indicate the disassembly and separation processes. End nodes that flow from these processes are labeled "recycle" or "scrap," and they indicate clumps of parts or materials that can be reused, recycled, or scrapped. In general, a smaller tree is an indication of a good design for recycle, especially if there is some demand for the clumps that are to be reused or recycled. Long diagram trees are less desirable because they indicate many levels of sequentially dependent operations. Constructing fish-bone diagrams forces designers to walk through a product's demanufacturing steps and to consider options for more efficient recycling.

Although the reverse fish-bone diagram is effective in improving the recyclability of a specific product model, it does not address the retirement process of product families and generations. Currently, one must construct a diagram for each product family and generation and compare them to see if a common retirement or demanufacturing facility can handle them all. Another shortcoming of the fish-bone diagram is that it fails to help engineers select materials and assembly designs that promote the recyclability of a product family.

The Recyclability Map

A recyclability map provides metrics that can help designers select appropriate materials and construct assemblies that improve recyclability. The metrics that are mapped include the following:

- **Variety complexity.** The total number of unique parts divided by the average number of parts in a product. A low number indicates commonality of parts in a product family.
- **Material complexity.** The number of types of materials used in a product.
- **Sort complexity.** The levels of the associated reverse fish-bone diagram and the number of clumps.

The total number of sort bins required for a product family retirement process is a good overall indicator of these three metrics. In general, the larger the number of bins the more complex the disassembly, the higher the material count, and the lower the uniformity of materials used. A good modular design for recyclability should have a minimum number of sort bins.

As evaluation measures are refined and evaluated, charts can be developed to help engineers understand the ratings for a product design and identify directions for improvement. Figure 3 plots the number of sort bins for an ink-jet printer against the percentage of parts or materials landfilled or incinerated. The example reveals two groups of subassemblies. Group B, at the bottom right of the chart, consists of parts that are mostly scrapped. Group A, represented on the upper left, includes the input/output (I/O) tray (or paper tray) that has a high recovery rate. Recyclability maps such as this one can help designers to prioritize areas where improvements can be made to enhance the reuse and recyclability of a product.

The recyclability map shown in Figure 3 was used by student teams at Stanford to redesign the paper tray. The improved design led to a reduction in the number of plastic materials from three to one, an improvement in disassembly resulting from a change in fastening methods, and a reduction in the number of

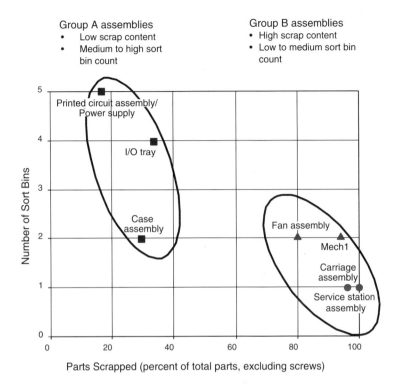

FIGURE 3 Recyclability map of an ink-jet printer.

sort bins from four to three. As shown in Figure 4, the reverse fish-bone diagram for the new paper tray has only one level. The redesigned paper tray also resulted in a 70 percent decrease in disassembly time (Figure 5) and a 60 percent reduction in scrap (from nearly 40 percent to less than 20 percent). Note that these ideas for improvement did not arise from constructing the reverse fish-bone diagram alone, but rather from the combination of the fish-bone diagram and the recyclability map, which helped the students target key areas of improvement.

The construction of the recyclability map for the ink-jet printer was possible only with detailed scrap rate information, which was obtained from the recycling plant. Where scrap rate information is not available, the rates can be estimated from material selection and recyclability data for each set of materials used in the design.

Scrap rates appear to depend on several factors, including the potential value derived from the reuse or remanufacture of the product, the material compatibility (valued either in terms of the mix of material used or the ease of separation of the material), and the value of the material that is recycled. These factors vary regionally and over time.

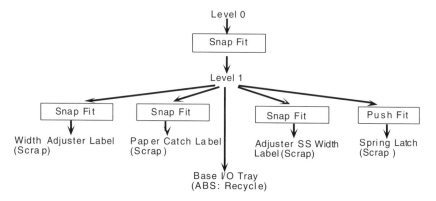

FIGURE 4 Simplified reverse fish-bone diagram.

Material compatibility and the value of the material recovered directly influence a designer's material selection for recyclability. This information should be part of a dynamic material database that also contains other environmental impact information such as energy use, resource depletion, and pollution effects. As part of its future efforts, the Stanford University Center for Professional Development plans to work on defining a material recyclability database, developing a method to predict scrap rates, and developing an Internet-accessible tool for constructing reverse fish-bone diagrams and recyclability maps.

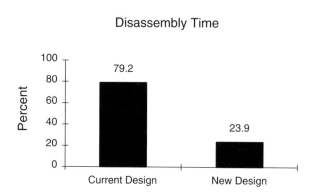

FIGURE 5 Reduction in disassembly time.

ACKNOWLEDGMENTS

The National Science Foundation Environmentally Conscious Manufacturing Program and the Lucent Industrial Ecology Fellowship funded the research discussed in this paper. In addition, the Stanford University Center for Professional Development supported the curriculum developed in the effort. Thanks also go to industrial partners Hewlett-Packard, Xerox, and Matsushita, as well as Burton Lee and the members of Stanford University's 1996 ME217 Hewlett-Packard project team who initiated the development of the recyclability map.

REFERENCES

Hauser, J.R., and D. Clausing. 1988. The house of quality. Harvard Business Review (May-June):63–73.

Ishii, K., C.F. Eubanks, and M. Marks. 1992. Evaluation methodology for post-manufacturing issues in life-cycle design. Concurrent Engineering: Research and Applications 1:61–68.

Ishii, K., and B. Lee. 1996. Reverse Fish-bone Diagram: A Tool in Aid of Design for Product Retirement. Paper presented at the American Society of Mechanical Engineers (ASME) Design Technical Conference, Irvine, Calif., August.

Marks, M., C. F. Eubanks, and K. Ishii. 1993. Life-cycle clumping of product designs for ownership and retirement. Pp. 83–90 in Proceedings of the ASME Design Theory and Methodology Conference, DE-Vol. 53. New York: ASME.

Sturges, R.H., and M. Kilani. 1992. Towards an integrated design for an assembly evaluation and reasoning system. Computer Aided Design 24(2):67–79.

Environmental Information in Supply-Chain Design and Coordination

PAUL R. KLEINDORFER and ELI M. SNIR

The past decade has seen wave upon wave of change in business operations and strategy. Fueled by the quality revolution in the early 1980s, companies began to see their business processes as the key focus of value creation. Business process improvement teams first worked locally on their processes and then on the quality of interconnecting links to upstream and downstream processes. Quality management evolved into time-based competition, then into process re-engineering, and finally into organizational transformation and the core competency movement. In all of this, process management remained a central focus for two reasons: first, because business processes provided a characterization of the business enterprise that enabled discourse at the strategic level; and second, because team-based approaches to continuous improvement found their best organizational match when assigned to well-defined processes.

Certainly, the two most important business processes identified during this period were the *supply chain* and the *new-product development process*. The supply chain consists of the subprocesses of procurement, production, distribution, and after-sales support. The extended supply chain is the cross-company supply chain resulting from linking suppliers and customers (and possibly suppliers' suppliers and customers' customers) to the supply chain of a particular manufacturing or service company. The new-product development process consists of the business processes, from research and development to product and process design to product launch activities that are required to design a new product or service and to maximize the likelihood that the product is a success in the market. Thousands of papers and books have been written in the interim about both supply-chain management and new-product development. By the mid-1990s, excellence in these two processes was viewed widely as a sine qua non for competitive success, with the speed, quality, customer focus, and cost of these processes being central aspects of both profitability and long-term strategy and competence.

At the same time as these revolutionary developments were occurring world-wide, a second and quieter revolution, sometimes referred to as *industrial ecol-ogy*,[1] was also occurring. To use the metaphor of the extended supply chain, the basic driver of industrial ecology was the notion that business itself exists within an extended supply chain of environmental and ecological resources, and waste and inefficiency in this ecosupply chain must be eliminated, just as economic waste must be eliminated in the narrower supply chain of business activities. Indeed, in the industrial ecology framework, each company has a special role as a steward of the environment and the ecosystem within which it operates. Natu-rally, this role of product stewardship and environmental waste and risk manage-ment encompasses both suppliers and customers, just as extended value-chain analysis encompassed suppliers and customers in the traditional supply-chain improvement process. And just as in the traditional model, it also has been found in industrial ecology that product and process design at the front end are better than end-of-pipe continuous improvement activities.

Our purpose in this paper is to review environmental stewardship activities in respect to the supply chain, with a focus on both design and continuous im-provement processes. We consider in particular the issue of how environmental information is gathered and used in these activities.[2] Our examples are primarily from the chemical and process industries, based on a series of interviews with leading companies in that industry connected to the Risk Management and Deci-sion Processes Center of The Wharton School at the University of Pennsylvania.

The concepts at the root of supply-chain design and improvement as a means of promoting sustainable development are not new. Life-cycle analysis, as a tool to understanding a product's complete impact on the environment, has been known for some time (White et al., 1995). This understanding is also widespread in the business community. Product stewardship, as promoted by the Responsible Care Code of the Chemical Manufacturers Association (CMA), emphasizes the need to encompass a product's entire span, from cradle to grave, to ensure envi-ronmental responsibility. Reverse logistics (Council of Logistics Management, 1993) is a related tool to reduce environmental impacts through recycling of packaging and products. Planning a product's prolonged life span, after normal use, also requires comprehending possible uses for used material.[3] Finally, the approval in 1996 of the first four of the International Organization for Standardization's (ISO) 14000 standards for environmental management systems can be expected to add significant further impetus to internal and extended supply-chain environmental stewardship activities.[4]

Information technology (IT) could play an important role in furthering the environmental and revenue benefits of such initiatives. Some examples of potential IT uses include gathering information on inputs and outputs of different processes, surveying customers and using prototypes to understand environmental impacts during product design, tracking product movements, optimizing transportation policies, and analyzing recycling and reuse behavior. With data from these sources, companies may improve the environmental aspects of their products at three important levels: (1) product and supply-chain design to minimize environmental impacts, (2) ongoing waste minimization and risk mitigation after the product has been deployed, and (3) diagnostic feedback from supply-chain participants to assess opportunities for new products and processes and to spawn future environmental initiatives.

These environmental improvements, in turn, can lead to economic benefits for companies in several areas: in helping customers to improve their performance and regulatory compliance; in reducing risk and strategic vulnerabilities internally and for their customers; in improving the atmosphere between themselves and regulators and possibly reducing compliance costs; and in improving their reputation and reducing transactions costs in dealing with local communities, environmental groups, and other external stakeholders. In all of these areas, careful tracking of environmental information (cost, value, and performance) is essential in understanding, managing, and legitimizing investments in product stewardship activities in the supply chain.[5] Let us explore these potential benefits in a bit more detail.

PROMOTING ENVIRONMENTAL EXCELLENCE

Corporate Image

Achieving environmental competence, and the use of IT to this end, is not only socially responsible behavior, it is good business. First, because of public concern about environmental issues, promoting environmental care can enhance a company's and an industry's image. Many companies are going out of their way to explain how they help environmental causes. The adoption of the Responsible Care Code by the CMA is a case in point. The chemical industry, led by the major chemical companies, is investing resources in and heavily advertising the importance of product stewardship to consumers and the public.

A major factor in a company's image is the occurrence of a major accident. Perhaps the most salient example of this is the chemical industry's reaction to the Union Carbide accident in Bhopal, India, in 1984. Not only did Union Carbide have to pay for remediation, it also lost much of its credibility as a responsible company. This led the company's reduction of the scope of its business through divestitures and closures by nearly 90 percent since the accident. The accident influenced many other companies in the industry and acted as a catalyst in reducing risks throughout the industry (Kunreuther and Bowman, 1997). In a recent assessment, Klassen and McLaughlin (1996), using event analysis, estimate that losses in shareholder value for large publicly traded firms from environmental incidents can be on the order of hundreds of millions of dollars per incident.

Regulatory Compliance

Regulatory compliance requires companies to track the use of hazardous substances and emissions of pollutants. Although actual compliance varies widely, especially among small firms with less to lose in the event of environmental incidents, major companies in the chemical and process industries have devoted significant resources over the past two decades to improving environmental performance (Jaffe et al., 1995). Indeed, many companies have begun to commit themselves to go beyond compliance. This is promoted by government agencies (e.g., in the XL and 33/50 programs)[6] that may apply different frequencies of enforcement and audit activities to companies with proven records. Such efforts also reduce the costs of changing technologies when new regulations are put in place because these regulations already will have been anticipated by companies with superior performance. A related benefit is the reduction of transactions and litigation costs when regulatory audits cite facilities. In ensuring compliance at the least cost, companies use state-of-the-art technologies, including IT, to ensure best practices. The role of IT in the compliance arena is in tracking current behavior, including wastes, releases, accidents, and "near misses," and designing new processes based on this information.

Liability and Negligence

Another factor driving companies to improve their environmental performance is the risk of being held liable, or found negligent, for accidents or environmental damage. When a company experiences an accident or an incident with significant real or perceived environmental damage, the company may be held liable to pay for remediation. This is true even when the company is acting prudently and using state-of-the-art technology. If corporate action does not meet social requirements, companies also may face punitive actions for their behavior. To limit liability and negligence claims, a company may choose to implement strict risk reduction mechanisms.

To this end, IT can be used to continuously evaluate the impact on the environment from industrial processes. This can be done by monitoring emissions from facilities, tracking the transportation of goods and services, collecting information on the effects of certain chemicals, and monitoring the use of products by customers and throughout the product life cycle. In general, the lower the level of pesticides, biocides, and toxics (PBT) associated with a company's products, the lower the level of risk from such liabilities. Thus, many companies are now moving to assess the PBT content of their products and are attempting to find substitutes that lower this content. Reducing PBT content is also an effective tool to reduce perceived risks by customers. As the use of potentially harmful inputs is reduced, so are claims of damage.

Supply-chain coordination plays a key role in limiting liability in three dimensions. First, to monitor a product's use throughout the supply chain, a company must be able to follow the product's flow. If a product is shipped to a customer who then uses it, perhaps mixing it with other products, and then passes it off to another party, it may be difficult to know how the product is used. If a client further down the supply chain uses the product and causes damage, the original producer still may be held liable. Proper design of supply chains is needed to eliminate such occurrences. Material Safety Data Sheets (MSDSs), mandated by the Occupational Safety and Health Administration (OSHA) for listed chemicals, have played an important role in explaining the proper use of chemicals in various production processes and ways to mitigate their environmental impacts (Baram et al., 1992).

Second, under the Comprehensive Environmental Response, Compensation, and Liability Act, companies face a joint and several liability standard. This means that even if only the final user of a hazardous substance disposes of it incorrectly, other companies that were involved in the process of production and distribution may be held liable to clean up the disposal site. This is especially problematic when the final user is unaware that a product may be hazardous and does not have the means to properly handle the substance. As might be expected, these provisions have led to a significant increase in supply-chain stewardship activity over the past decade (Snir, 2001).

Third, reducing liability in the supply chain can be achieved by not producing certain substances that have a high probability of being misused, or by choosing responsible supply-chain partners (Snir, 2001). Both of these tactics were evident in the product stewardship activities of the companies that we interviewed.[7] Note that changing distribution channels or supply-chain partners does not guarantee a company less liability, and it has additional disadvantages. Liability may not be reduced if risky customers still purchase the company's products through intermediaries (e.g., distributors or formulators). Thus, products still may be sold to unsophisticated customers, but the focal company does not have the means to train those customers in proper material handling and use. A further disadvantage is the increased distance between a company and its customers. In dealing through distributors, a company may not get enough feedback from its products' end users and may not be able to respond to their needs. In this case (which is typical of agricultural products), product labeling and other communication media, including the Internet and trade association publications (e.g., to crop growers), are being used increasingly to improve stewardship.[8] IT can be important in tracking awareness of product use and handling procedures, especially if coupled with electronic commerce methods (using direct or third-party fulfillment procedures) of ordering the product. Nevertheless, when multiple organizations are involved in the manufacture or distribution of products, environmental stewardship remains a difficult problem.

Another important factor in the use of IT in reducing environmental liabilities is the ability to set industry standards and jointly define best practices. When there are common practices throughout an industry, adhering to these practices is a prima facie defense against negligence. This philosophy leads companies to share knowledge of substances and possible outcomes, set joint guidelines for use, and promote the use of industry best practices. However, such guidelines are typical only for commodity products or generic handling practices, because information (even on handling and use) for more proprietary products is usually quite sensitive and is distributed only to customers.

Community Relations

Improved relations with local communities and other external stakeholders are becoming increasingly important for companies, as a matter of both law and of best practice (McNulty et al., 1998). This has led companies to improve their environmental practices and the information they make available to the public concerning these practices. By doing so, a company can maintain its social franchise while enhancing its economic franchise. In communities, public-interest groups are increasingly pressuring companies to practice environmental prudence and to prove those actions. The availability of easily accessible information on a company's environmental objectives and measurable performance criteria therefore can be expected to play an increasingly important role in assuring stakeholders that a company or facility is adhering to stated objectives.[9]

A frequently cited example of the interaction of information and performance is in the area of Toxics Release Inventory (TRI) reports. Parallel to the requirement to file TRI reports under the Superfund Amendments and Reauthorization Act Title III in 1986, many companies formed community advisory councils that played an important part in reducing toxic emissions. The fact that stakeholders had access to information regarding emissions, and the fact that this information was required by law, led companies and their communities to take action to reduce emissions. Accurate databases play an important role in such a process. They allow a company to monitor its own activities against industry averages and they allow external stakeholders access to (and influence on) a company's environmental performance.[10]

A further example of the growing importance of public information systems related to a company's environmental performance is the requirement, under section 112(r) of the Clean Air Act Amendments of 1990, that risk management plans (RMPs) be developed by a number of facilities that store certain chemicals and that summaries of these RMPs be made available to the public. Doing so will surely increase the importance of IT to companies and their supply-chain partners in ensuring that these plans are kept up to date and are responsive to public concerns.[11]

Another important area in which companies can enhance their reputation in the community is waste reduction. This pertains both to hazardous wastes and to other industrial wastes. In the United States, landfills and other means of disposal are becoming a large environmental problem, affecting many communities. Companies will be required by local communities to do their share in reducing waste. This can be done at all levels of product development and production: by designing for frequent reuse and recycling, by producing with less waste and fewer emissions, by reducing transportation and packaging, and by promoting recycling in local communities. Many of these measures also reduce costs because they require using less virgin resources and production inputs. Companies such as Procter & Gamble (White et al., 1995; Hindle et al., 1996) are making plain in their marketing and promotion programs, as well as in their internal practices, their commitment to producing "more with less." IT plays an important role in waste reduction by monitoring the inputs and outputs of every stage of the supply chain and by providing key input to the design of future products and packaging.

Employee Health and Safety

Similar to community concerns, employee health and safety (H&S) is a key focus of product stewardship. It is promoted both by general duty clause requirements of prudent management as well as by specific requirements, such as the Process Safety Management Standard under OSHA. Employee H&S is not limited to company workers or on-site exposure, but includes all parties in the supply chain who may be exposed to a company's products. Often, employees are the first victims of poor safety standards that later evolve to environmental dangers. Emphasis on responsible behavior in respect to employee H&S may truncate such processes before they develop, allowing timely and cost-effective measures to be taken.

Emphasis on employee H&S also has direct effects on output and productivity. Improved employee health reduces costs associated with sick leave and health insurance. Worker relations may also be improved when health concerns are minimal, and safer conditions can boost employee morale, ultimately leading to greater productivity. These factors should be true both for the companies that promote product stewardship and for their clients.

Customer Relations

Gathering information about a company's product and its uses allows for improved relationships with vendors and customers. A deeper understanding of a product's uses and benefits promotes product innovation (von Hippel, 1988) and can lead to improved designs that minimize waste and unnecessary procedures.

Such understanding also enables companies to offer their customers suggestions on ways to reduce their emissions or product liability. In cases where accident prevention is of interest (both on-site and to customers), learning the causes of accidents enables companies to take steps to reduce risks. To this end, MSDSs are helpful in explaining proper usage. Understanding product usage also helps in eliminating waste by encouraging source reduction and reuse (Willums and Goluke, 1992). These means of increasing a product's user-friendliness depend on close ties with customers, so that product stewardship is not just a means of avoiding liability and ensuring regulatory compliance, but is also a business driver of product and process innovation for all supply-chain participants.[12]

Economic Motivation

As noted above, several important tools have emerged in the past decade to promote stewardship in the supply chain and overall economic efficiency in the extended supply chain. These tools include life-cycle analysis, reverse logistics, and several gated "design for X" screens (where X includes factors such as environment, safety, disassembly, and recycling). Each of these tools is directed toward two complementary drivers of economic value in environmentally sensitive activities: (1) measurement and assessment of environmental impacts throughout the supply chain and (2) reduction of either impacts or capital and operating costs by product and process innovation.

Life-cycle analysis assists in identifying the sources of waste and pollution from cradle to grave (Dillon, 1994). One advantage is the reduction of excess inputs and wastes throughout the supply chain, thereby lowering costs and promoting sustainable development. Other benefits include reducing accident risks and lowering emissions, each of which leads to lower costs in the long run.

Reverse logistics is an additional tool to reduce costs, through recycling and reduced source inputs (see Box 1 for an example). By understanding a product's use throughout its extended life cycle, a company may find ways to design modules for reuse, recycling, reclaiming, and reconfiguring. This typically requires modular design from the beginning of product design, which also enables maintenance and upgrading at different stages. Such design for disassembly, on and off site, is becoming widespread and allows for increased use of reverse logistics (Council of Logistics Management, 1993). Reverse logistics is especially well known for its applications in the reuse of packaging. Instead of single-use boxes, many companies are turning to nondisposable packaging to improve environmental impact and costs. One commercial example is the reuse of wooden pallets between distributors and clients. In the transportation of hazardous materials, reuse is especially important. IT plays an important role in tracking container location, materials within containers, and verifying that containers are not filled with substances that may react with residue.

Reduced costs also are realized through reduced transportation usage.

BOX 1
Reverse Logistics at DuPont: The Petretec Case

An illuminating reverse logistics initiative, born of the sustainable development philosophy at DuPont, is the methanolysis of polymers and plastics, popularly termed unzipping polymers, to recover near-virgin material. Petretec, the company's polyester regeneration technology, unzips film, fiber, and plastics to their raw materials, dimethyl terephthalate (DMT) and ethylene glycol (E6), which then are used instead of virgin resources.[1] This allows DuPont to reduce both the use of oil-derived feedstocks and the amount of waste buried in landfills. Currently, with shortages of DMT, the ability to recycle polyester materials is extremely worthwhile.

DuPont invested more than $12 million to convert an existing DMT production unit to the Petretec facility near Wilmington, North Carolina. The facility should have the capacity to produce 100 million pounds of DMT and 30 million pounds of E6 per year. The process is extremely robust, and it can accept polyester with a variety of contaminants at higher levels than acceptable in other processes.

An important market for polyester is auto manufacturing. With the growing awareness of design for disassembly and design for environment, technologies that offer to reduce the environmental impact of cars seem promising. Research is under way to produce cars that have a large percentage of recyclable material. With Petretec there is a new interest in attempting to design the entire interior of a car with polyester materials. This will allow car manufacturers both to reduce the costs of building cars and to increase the ease with which cars can be recycled.

Another important market for recycled polyester is food containers. The Petretec process is approved by the Food and Drug Administration, so a syrup bottle can become a computer tape, then an x ray, then a videotape, then a popcorn bag, then an overhead transparency, then a polyester peanut butter jar, then a snack food wrapper, then a roll of film, and then a syrup bottle again.

---

[1]For an early description of recycling in DuPont's film division, see Council of Logistics Management (1993). The Petretec is described at http://www.dupont.com/corp/ products/dupontmag/novdec96/pertec.html [October 20, 1998].

Optimizing the needs to transport material, or using more energy-efficient means, promotes environmental prudence and lower costs. In the logistics domain, transportation is a crucial environmental factor, accounting for 11 percent of U.S. expenditures for goods and services and 25 percent of recycling costs (Wu and Dunn, 1995). Promoting improved environmental transportation can have a positive impact on the bottom line. For example, transportation costs may be reduced by using efficient loading, scheduling, and routing techniques, including consolidation of freight and balancing of backhaul movements. IT obviously plays an important role in any attempt to minimize the environmental effects of product transportation.

Another cost-effective means of reducing transportation-associated environmental effects is the use of alternative fuels. The U.S. government is taking a strong position for this cause. The Energy Policy Act of 1992 requires that, by 1998, 50 percent of all purchased federal cars must be powered by alternative fuels. One of the alternative fuels being examined by the U.S. Postal Service is compressed natural gas, which is 40 percent cheaper than gasoline and has a substantially lower total impact on the environment. Another example of the use of alternative fuels is the testing of liquefied natural gas by Union Pacific Railroad to power its locomotives. Again, the driving force behind this transformation is the need for cheaper, safer, and cleaner sources of fuel (Wu and Dunn, 1995).

Environmental prudence throughout the supply chain lowers costs not only for the company, but also for customers and vendors. Reduced production and logistics costs allow for lower costs to customers, but this is not the only means of lowering partners' costs. With life-cycle analysis, all processes that a product goes through are examined, and more efficient processes[13] are developed, regardless of their position in the supply chain. This is especially true for source and waste reduction techniques used in life-cycle analysis. When new processes are developed that minimize waste, often the final customers are those that benefit most. Some of these benefits are derived from less waste being generated, but lowering emissions also gains customer approval.

Enhanced use of reverse logistics also reduces costs for customers. Companies may offer different prices for modules that are reused, which may be a major advantage to certain customers. Promoting reverse logistics reduces on-site waste for customers and relieves them of the need to process such waste. With hazardous waste, this benefit becomes extremely important for some customers who do not have the means or the desire to deal with such waste. Ashland Chemical, a large chemical producer and distributor, is offering more services to customers who want to minimize the amounts of chemicals on site (Chemical Week, 1991).

DRIVERS OF ECONOMIC VALUE AND ENVIRONMENTAL EXCELLENCE IN A SUPPLY CHAIN

Figures 1 and 2 show the internal and extended supply chains, respectively, in relation to the drivers of economic value and environmental excellence noted in the above discussion. The key insight derived over the past decade is that the supply chain, from materials procurement to manufacturing to logistics to recycling and disposal, should be viewed holistically where the environment is concerned. Each stage gives rise to its own effects, impacts, and opportunities for improvement, but effective environmental strategies require an analysis that encompasses the entire supply chain. This not only reduces sources of risk and liability by reducing pollution, wastes, and hazards, it also promotes reduced costs and better products. This expanded view of product stewardship and supply-chain management is gradually transplanting the traditional view focused on internal environmental excellence and caveat emptor. In addition to the above-mentioned specific tools to promote this emerging concept of product stewardship, an important

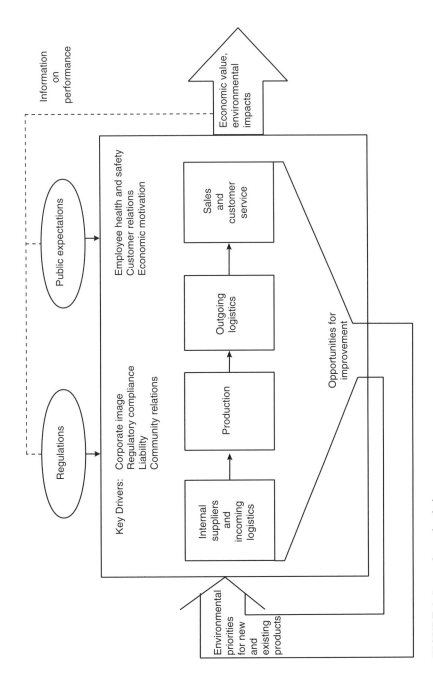

FIGURE 1 Internal supply chain.

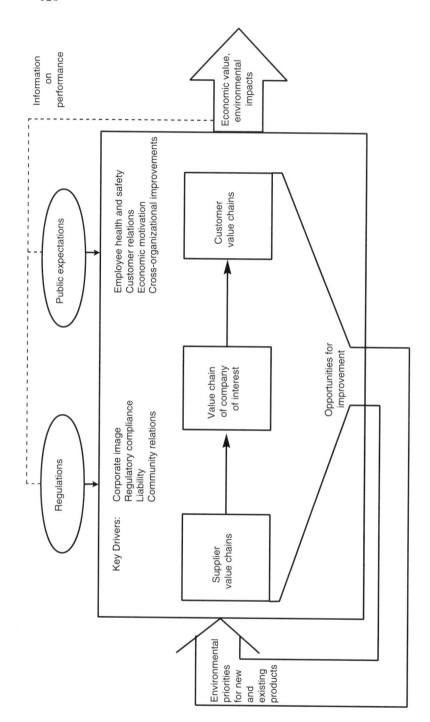

FIGURE 2 Extended supply chain.

general capability is effective and efficient IT in the design and coordination of supply chains and as a feedback mechanism to further improve diagnosis and performance. We now consider in more detail the use and management of environmental information to add value in the extended supply chain.

ENVIRONMENTAL INFORMATION IN SUPPLY CHAINS

We now understand that the design stage of products and processes determines the major consequences of these, including a large part of a product's environmental impact (Ulrich and Eppinger, 1995). To ensure that proper care is taken, life-cycle analysis and other techniques may be useful. In this analysis, environmental information plays a key role in a number of dimensions including source reduction, transportation optimization, emission analysis, and reverse logistics.

Source Reduction

Utilizing environmental information to reduce the use of inputs can be accomplished as a part of material balances incorporated in life-cycle analysis.[14] Collecting information about risks from similar products and processes is important in comprehending the environmental impact of a new process. Coupled with simulation of alternative options for product and supply-chain design (White et al., 1995), the outcome of this analysis is products and processes that minimize the use of raw materials.

The business process for accomplishing source reduction and life-cycle analysis consists of overlaying the new product development process with a series of screens that subject new products and processes to a detailed assessment. Allenby (1994) states that information should be collected on at least four dimensions in analyzing the life-cycle environmental impacts of a product. These include environmental (ecosystem), manufacturing, social/political, and toxicity/ exposure (human) impacts. Information on these impacts then is coupled with the company-specific, multiphase product development process. This coupling ensures that environmental considerations are taken into account in addition to customer demand, manufacturing processes, engineering design, and profitability. This method, often called design for environment, allows for environmental factors throughout a product's life cycle to be assessed at the design stage. The primary type of IT used in the design stage is the database with information regarding the uses of different materials. These databases include information on the hazards of certain materials, their toxicological properties, and other relevant environmental information. With the active assistance and participation of safety, health, and environment (SHE) experts and product stewardship representatives on the new-product development team, the indicated multiphase approval process helps to minimize PBT content of new products as well as to develop, during the design phase, an extended supply-chain perspective on the traditional economic impacts of products as well as their broader ecological impacts.

Transportation Optimization

In optimizing transportation methods at the design stage, IT plays an important role. Deciding at which stage to implement certain production processes determines the amount and types of materials to be transported. The analysis of energy and environmental intensity of alternative supply-chain designs is in its infancy, but this can be expected to grow rapidly as sustainable management practices take root (Hart, 1997), especially if concerns about global warming intensify. Minimizing energy intensity and environmental impacts of alternative supply-chain designs is, in principle, a straightforward simulation exercise if data are available. Doing something about these impacts requires that this analysis be undertaken at the design stage.

In addition to using IT to determine optimal transportation schemes, the design stage requires development of IT methods that will assist in coordinating the movements within the supply chain. This is true for forward logistics, such as product tracking, as well as for reverse logistics. Bar-coding of containers, for example, is important for tracking containers and materials throughout the supply chain. In addition to the recognized benefits of vehicle routing and replenishment improvements that such information can enable (Fisher et al., 1983), container sensors can provide telemonitoring of contents, pressure, and temperature, which is increasingly important for both improved order fulfillment as well as product stewardship assurance. At present, however, such information is gathered primarily for business purposes, and its use for the environmental assessment of alternative transportation and distribution systems is secondary.[15]

Emission Analysis

Emission analysis is an important part of supply-chain design. Different production processes emit wastes using different media such as air, water, or land (which also are regulated under quite different laws and regulatory standards, although the current Sectoral Initiative of the U.S. Environmental Protection Agency (EPA) may bring some changes to this regime). Also, these emissions are at different physical locations and under the responsibility of different parties in the supply chain.

IT plays two major roles in the design stage with regard to emission analysis. First, information is needed to understand how certain processes affect risks and emissions, as was stated in the discussion of life-cycle analysis. Second, means to monitor emissions must be put in place while designing processes throughout the supply chain. These are important for material balances ex-post. Measuring inputs and outputs of a process are crucial for those who wish to validate hypotheses concerning the actual behavior of processes. Currently, industry environmental leaders are quite proactive in reducing emissions, with specific, measurable targets set for each business unit and each facility. These include emission reductions as well as recycling and reuse. Increasingly, these are being used by senior management to review progress toward better environmental practices as well as reducing underlying drivers of cost and risk in a company's businesses. IT is clearly a foundation for all of this activity.

Reverse Logistics

A final use for IT in the design stage is in the reverse logistics aspects of the supply chain to design the entire product life cycle from cradle to grave. Tracking the location of products and packaging is an important part of a reverse logistics network. Reuse of packaging is an important and cost-effective means of reducing environmental impact. To fully utilize this method, tracking of packaging location is important. This may be used to optimize backhaul routes or to optimize the transportation of hazardous materials. Tracking products is crucial in optimizing the reverse logistics process. In many areas, reused or recycled materials are, or are perceived to be, of lower quality than new items, and companies must be able to differentiate between them. This can be done by bar-coding products or implanting invisible footprints. Data on the number of reuse cycles that a product has gone through may influence product quality, perception, or price. For example, BMW has a longstanding policy to reuse and recycle parts from old cars. To expedite the process they code each recyclable part (Wu and Dunn, 1995).

Coordination of the supply chain is facilitated by electronic data interchange (EDI). This allows different parties in the supply chain to gain knowledge about product use, product and packaging location, emissions, stock on hand, and customer use. Product use can be facilitated by making MSDSs electronically available. Customers access the MSDSs (and other just-in-time learning instructions) to ensure the safe and efficient use of a company's products. Stock on hand at different sites or at customer locations is, of course, also beneficial for optimizing shipments of material. The great advantage of EDI is that it greatly reduces the costs of information transfer and allows multiple parties, including those with product stewardship responsibilities, access to all relevant data.

Feedback from Information Technology Systems

Another important advantage of implementing environmentally oriented IT systems throughout the supply chain is the ability to gain feedback concerning actual behavior to further improve the supply chain. This information may be useful in product environmental audits, ex-post investigations of accidents, reviews of near misses, establishing guidelines and verifying their adherence, and for product stewardship review boards (PSRBs). Clearly, for audits concerning a product's environmental impact, actual information on all dimensions of environmental impact is key. Without such information, auditors may only examine the process design and hypothesize about whether original targets were met. Using actual data, original targets can be assessed, assisting to establish improved processes and to validate engineering designs. Moreover, in case of accidents and near misses, viable information is key to learning what went wrong and how to correct it.[16] Similar to the use of black boxes in airplane accidents, solid information about processes and emissions is key to unraveling cause and effect.

In addition to the business performance uses of environmental IT noted above, IT is also central in allowing senior management to measure progress for its employees, investors, and external stakeholders in achieving environmental improvements. Many companies have utilized IT to track and reduce their TRI and other key environmental indicators over the past few years. For example, Eastman Kodak's top management has set specific performance goals, which include employee health issues and assessment of environmental responsibility. The company also monitors other environmental impacts and measures its improvement over time against targeted commitments.[17] Similarly, the Brewers of Ontario publicly committed in 1991 to recover and reuse 100 percent of its sold beer packaging. Currently, they recover 99 percent of all bottles, 83 percent of all beer cans sold, and 98.4 percent of all beer packaging.[18]

Finally, PSRBs are one means whereby senior management can establish leverage and influence the company's handling of environmental issues. The PSRB is established in accordance with the Responsible Care Code to periodically review a process or product. Major responsibilities of the PSRB are to assist line management in product stewardship and to reduce the environmental impact of the company's products (Bond, 1995). Coupled with an overall environmental policy review board as part of the executive committee, the PSRB can be a powerful instrument for ensuring that environmental commitments are embodied in the new-product development process and in the operation of the extended supply chain.

MANAGERIAL SYSTEMS USED TO DRIVE
ENVIRONMENTAL EXCELLENCE

Integrating the supply chain to ensure environmental excellence requires integration with key business processes, measurement of results, and commitment from top management. A number of managerial concepts exist that promote these steps toward environmental prudence; collectively, they are called environmental management systems (EMSs). Of these, the best known are EMS structures embodied in the newly approved ISO 14000 standards and the CMA's Responsible Care Program. These managerial systems require IT to ensure that managerial goals are being met, but both leave considerable discretion to each company to decide how their EMSs will be structured. We review here the ISO 14000 standards and their relationship to supply-chain environmental information.

ISO 14000 is actually a series of environmental management standards. These standards are voluntary, and, taken together, they provide guidelines for the development and maintenance of an overall management system, designed to meet individual company needs, but comporting with general requirements for effective environmental management. The standards themselves were written by international cooperating industrial groups and government environmental and standards organizations under the general guidance of ISO, a private-sector, international standards body based in Geneva. Founded in 1947, ISO promotes international harmonization and development of manufacturing, product, and communication standards. The closest relative to the ISO 14000 standards are the ISO 9000 standards for quality. ISO 9000 was set up as a management system standard in the 1980s and spread rapidly throughout the world, as organizations found that standardization of documentation, training, and data structures for quality could promote significant improvements not only within the boundaries of a single organization but across national and international supply chains. ISO 14000 began development in 1991, after the successful deployment of ISO 9000 standards, and the aspirations underlying ISO 14000 were motivated by the experience with ISO 9000. Indeed, many organizations recognize synergy between ISO 9000 and ISO 14000, and they hope to achieve superior environmental performance by extending their ISO 9000 experience and management systems to incorporate additional environmental features required by ISO 14000.[19]

As foreseen in ISO 14000, EMSs are management system standards for process guidelines, not performance standards. EMSs help an organization to establish policies and meet its own environmental objectives through documented accountability and responsibility structures, communication and training programs, and management control and review functions. Companies may choose to be certified for either specific facilities or for the company or division as a whole. EMSs may not set specific requirements for environmental compliance, but they do call for a commitment to compliance with environmental laws, prevention of pollution, and continual improvement of environmental performance. EMSs can include specific compliance statements and procedures, and these can be audited as part of the ISO 14000 EMS certification process. Thus, ISO 14000 could provide additional assurance of compliance with those laws and regulations with which the EMS asserts compliance. The following standards are the initial standards foreseen in the ISO 14000 series[20]:

- 14001: EMSs—specification with guidance for use.
- 14004: EMSs—general guidelines on principles, systems, and supporting techniques.

- 14010-12: Principles, qualification criteria, and procedures for internal and external auditing.
- 14014: Initial review guideline to determine a corporation's baseline operating position, typically used prior to establishing EMSs.
- 14031: Guidance for measuring environmental performance over time.
- 14020-22: Describes labeling principles such as self-declarations of environmental benefits of products.
- 14040-43: Establishes a methodology for a product's life cycle, including assessment impacts and improvement analysis.
- 14050: Terms and definitions.

A closely related set of requirements and standards is that promulgated by the European Union (EU) under the Eco-Management Audit Scheme (EMAS). Like ISO 14000, EMAS is a voluntary program to promote continual environmental improvement in the private sector. EMAS identifies for the public and publishes every six months the names of those companies that meet EMAS standards. Companies meeting these standards may place an EU-approved logo and statement in their publications and letterhead. EMAS became operational for participation in April 1995. As currently implemented, EMAS has additional and more stringent requirements than ISO 14000, including the requirement that the certification statement itself, as well as containing specific information verifying continual performance improvement, be made public. Note that although the ISO 14000 standards require a commitment to continual improvement, at least in the company's environmental management systems, they do not require a verification of continual improvement in environmental performance. Current plans call for the EU to reconcile EMAS and ISO 14000 by accepting ISO guidelines with an explanatory document specifying the additional EMAS requirements. However, the details of whether and how EMAS and ISO certification eventually will be reconciled are not yet resolved.

One final point is essential. The auditing requirement for ISO 14000 can be executed by either the organization itself (an internal audit) or by a qualified third party. Requirements for auditors are spelled out in ISO 14010-12. What one can expect to occur is that third-party external auditors will become the vehicle of choice, because of the added element of objectivity of third parties as well as for economies of scale in performing the audit function and related value-added services provided by third parties.[21] It is sometimes noted that small and medium-size companies, in particular, will want to undertake internal auditing procedures rather than hire external auditors. In our view, this is very unlikely to be the case. Either external auditors will add sufficient value to make their services worthwhile or small companies will not find it useful to become ISO 14000 certified in the first place. If, as was the case with ISO 9000, smaller companies become certified in order to satisfy larger customers downstream in the supply chain, then such customers will almost certainly require an external audit to verify ISO 14000 compliance. Where there are sector-specific benefits (e.g., chemical distribution, dry cleaning), and small to medium-size firms are involved, one would expect trade organizations to help standardize generic EMSs to capture best practices and to ensure synergy with the regulatory process. Such generic EMSs would again logically be in the hands of external service organizations to deploy and to audit.

Concerning the structure of implementation, ISO 9000 provides a reasonable model of what to expect. Third-party organizations, including consulting and auditing service companies, will provide services to assist companies to prepare for certification and as well as providing auditing and certification services. Business and trade organizations can be expected to play a major role in establishing generic EMSs and in assisting certification organizations to understand value-adding services that may accompany ISO 14000 certification or auditing. State and federal agencies may have responsibilities for licensing qualified auditors and auditing organizations and for continuing to monitor compliance with applicable laws and regulations. If ISO 14000 is an efficient way of improving compliance and performance, then one would expect it to become a global standard, just as ISO 9000 has in the quality arena, and to drive the very vision and structure of what constitutes effective EMSs and practice. Whether this will happen clearly depends on the balance of costs and benefits of ISO 14000 relative to other methods and systems for achieving effective environmental performance. The potential benefits from ISO 14000 stem in part from the commonality of practice that standards are intended to promote, together with improvements in both cost and performance. For our purposes here, we note only that ISO 14000 is being increasingly viewed as a potential standardized vehicle for structuring and auditing EMSs, both across business units in a given corporation as well as across the extended supply chain. Just as in ISO 9000, the promise here is that standardized practices will establish a discipline, understandable across organizations and business units, for identifying opportunities for environmental improvement and monitoring against agreed performance metrics and targets. Whether this promise will materialize for ISO 14000 remains a central open research question.

SOME RESEARCH QUESTIONS CONCERNING ENVIRONMENTAL INFORMATION IN THE SUPPLY CHAIN

The above analysis of environmental information in supply-chain design and coordination highlights a number of interesting questions that warrant further research. These questions are posed in a manner derived from our interviews with representatives of the chemical and process industries. They may be interpreted both as key questions of practitioners in the middle of the continuing revolution of industrial ecology and as the usual academic end-of-paper positioning for future funded research.

Use of Environmental Information

The first question of interest regards the actual use of environmental information in product and supply-chain design. For various industries, what is best practice and good practice regarding the assessment of a product's or supply chain's environmental impact during the design stage? What models or templates are used and how product- or process-specific are they? In particular, what criteria are important for both internal and external assessment of environmental impacts? Finally, is environmental information and supporting IT an add-on or is it fundamental to the new product or process development and design process?

Investment Drivers

A second question of interest concerns the drivers of investments in product stewardship and environmental excellence in the extended supply chain. Are these primarily in ensuring compliance and reducing internal risk (e.g., through reducing PBT content of products), or are they forward-looking, value-added aspects, with full environmental considerations, such as improved yield, reduced energy input, and improved customer loyalty, essential drivers? In a word, among the various elements of potential benefit noted in this paper, where is the "money" in practice resulting from product stewardship for the extended supply chain? In particular, do other partners in the supply chain fully appreciate initiatives taken on by one of their supply-chain partners? For example, it might be hypothesized that customers would be willing to reward a supplier in several ways for its environmental leadership and stewardship activities. These include paying higher prices for goods, giving a higher preference to the supplier for repeat business, and cooperating with the supplier in the development of new products or services. It would be of considerable interest to analyze in terms of rewards such as these what the returns are to customer support activities in the environmental area.

Environmental leadership requires the support of top management. It also requires coordinating between different business units and different companies, within the supply chain. The coordination requirement would seem to lead to SHE activities being organized at corporate headquarters. On the other hand, organizational transformation and "flattening" in the 1990s have witnessed the mainstreaming of SHE activities from headquarters to the different business units. But many companies believe that maintaining the capability to launch new products and to monitor and lead SHE activities for existing products requires a continuing strong corporate presence. What factors drive the balance here? Are there facilitating mechanisms to overcome the coordination and monitoring costs arising from mainstreaming? What role can and will ISO 14000 and supporting IT play, both internally and across the supply chain, in facilitating this balance?

Performance Metrics

Performance metrics should be a continuing focus of research. Under ISO 14000, such performance metrics, and the definition of business processes, will provide the architecture against which audits will be conducted, with monitoring, learning, and mitigation activities triggered by these. If the area of quality management is any guide, and we think it is, what will be treasured is what gets measured. In particular, causing each business unit to analyze current PBT content in their products and other specific product and process environmental indicators will cause business units to review these indicators and to move in the desired direction, just as tracking TRI data and OSHA reportables have "caused" fundamental changes in the underlying processes giving rise to these data. Several questions arise here. What structure of environmental performance metrics is useful for management control? How can such metrics be made more visible to those who affect the outcomes? How should these be coordinated with environmental strategy for the business units and with various key business processes implementing this strategy (e.g., product and supply-chain design, product stewardship, regulatory compliance, customer support, community and investor relations)?

Possible Downsides

A final research question raises the issue of potential adverse implications of environmental information. Not all aspects of producing and disseminating environmental information are positive for the company producing that information. A company's environmental behavior may be of interest to other companies in the supply chain, competitors, or external stakeholders including regulators, communities, and public-interest groups. Providing raw data may have some unwanted effects, such as increased vulnerability to liabilities (including provable negligence) and revealing technologies and performance to competitors. What is the appropriate balance between the trust-promoting and emergency planning benefits of environmental information to external parties and the costs of providing this information and coping with the impact of use and misuse of this information by external parties?

ACKNOWLEDGMENTS

Many of the results of this paper are derived from roundtables and interviews supported by a cooperative agreement from the EPA to the Wharton Center for Risk Management and Decision Processes. Their support is gratefully acknowledged. Helpful discussions with William Lorenz, Irv Rosenthal, Stan Schechter, and Ernest Weiler also are acknowledged.

NOTES

[1]For an introduction and overview of industrial ecology, see Allenby and Richards (1994). For a discussion of international trends in industrial ecology, see Munasinghe (1996) and Hindle et al. (1996).

[2]A very similar analysis applies to safety and health considerations, and, indeed, product stewardship is directed toward mitigating all such impacts. We focus here on environmental issues because they are typically more difficult to align with business value-added than health and safety issues. But much of our discussion applies equally to these latter issues as well.

[3]Willums and Goluke (1992) provide background and case studies on extended product life.

[4]Kleindorfer (1997) provides some background on the expected scope and impact of ISO 14000.

[5]As a senior executive at a major chemical company noted to the authors, economic drivers at the business unit level are central in promoting product stewardship throughout the supply chain. Every initiative must be legitimized either as "required compliance" or adding value for the company or its customers. Many initiatives currently under way stem from customer demands and are the direct result of product stewardship activities. As business units develop a philosophy of improving environmental impacts throughout the supply chain, they become proactive in promoting changes that reduce environmental impacts for supply-chain partners. Many of these result in direct economic benefits as well, for example, through energy or material savings or through recycling benefits.

[6]The U.S. Environmental Protection Agency's (EPA) 33/50 Program (also known as the Industrial Toxics Project) is a voluntary pollution reduction initiative that targets releases and off-site transfers of 17 high-priority toxic chemicals. Its name is derived from its overall national goals—an interim goal of 33 percent reduction by 1992, and an ultimate goal of a 50 percent reduction by 1995, with 1988 being established as the baseline year. The 17 chemicals are from the Toxic Release Inventory. They were selected on the basis that they are produced in large quantities and subsequently released to the environment in large quantities; they are generally considered to be very toxic or hazardous; and the technology exists to reduce releases of these chemicals through pollution prevention or other means.

The EPA's Project XL was created under President Clinton's Reinventing Environmental Regulation Initiative. Project XL involves the granting of regulatory flexibility in exchange for an enforceable commitment by the regulated entity to achieve better environmental results than would have been attained through full compliance with the existing regulation.

[7]As one executive put it, "We have a very stringent screen for new products. We will not deploy a new product which we believe has a significant probability of causing environmental harm through mishandling or misuse, even though correct use of the product would not be harmful. In addition, in recent years [the company] has changed its distribution channels, relying more on large customers, who have a track record as being responsible. Riskier supply-chain partners have been dropped due to the possibility of accidents or product misuse on their part; we just can't afford this."

[8]Further evidence is the fact that many companies, in diverse industries, now publicize their environmental policies on the Internet. The list of such companies includes DuPont, Procter & Gamble, Brewers of Ontario, Eastman Kodak, Rohm and Haas, and Exxon. Each company emphasizes its responsibility to the environment and the importance of environmental factors in its decision making, and many companies even inform the public of specific targets they have for reducing environmental impact. For example, Rohm and Haas publishes the findings of the Responsible Care Management Systems Verification conducted in 1996 under the guidance of the CMA. (See http://www. rohmhaas. com/company/msv.html [October 20, 1998].) We can expect such public outreach activities to increase and, with them, the need for better environmental IT systems to allow information of this type to be produced and accessed economically. For this to happen, such information increasingly is being collected as a routine part of business activity and not just for public relations purposes.

[9]In this regard, the experience in Europe with the Eco-Management Audit Scheme and the growing experience worldwide with ISO 14000 are cases in point. Environmentalists have pressed very hard

to see both of these standards include measurable performance results, not just general objectives. The balance between guarding competitive information and informing the public has yet to be worked out in detail, especially in ISO 14000, but this surely will remain an important focus in discussing the impact and acceptability of these standards by environmental groups. For further discussion, see Kleindorfer (1997).

[10]For further discussion of the important role of information as a means of increasing the efficiency of environmental regulation and performance, see Kleindorfer and Orts (1998).

[11]For further discussion of section 112(r) and the role of information in its implementation, see Rosenthal and Theiler (1998).

[12]In the interviews surrounding this study, however, most companies indicated that they were still being driven by a liability and compliance mindset in product stewardship. Except in the safety domain, the value-adding potential of stewardship activities remains largely to be exploited in both design and improvement opportunities.

[13]Efficiency here is with respect to the constrained cost minimization problem that the company faces, that is, minimizing total life-cycle costs while taking into account current and future regulatory and public policy constraints.

[14]For an interesting application of material balances on environmental impact reduction, see Ayers (1997).

[15]Indeed, in interviews with several chemical manufacturing and distribution companies, it is clear that the analysis of transportation alternatives currently is driven primarily by business factors such as cost and time, with safety in handling and routing also a consideration, but with very little explicit concern for environmental impacts. This could change quickly if control of nitrous oxides and volatile organic compounds related to ground-level ozone mitigation becomes more costly, or if carbon dioxide suppression related to global warming becomes an important priority.

[16]In any case, such information will be required as part of the accident history database that all regulated facilities are required to file under section 112(r) of the Clean Air Act Amendments, which also will require the first five-year history to be filed by June 1999. In the United States, it is estimated (Rosenthal and Theiler, 1998) that there are some 66,000 facilities covered under this rule, so one can soon expect environmental information systems with the capabilities to record not only releases but accident histories to be standard practice.

[17]See http://www.kodak.com/aboutKodak/corpInfo/environment/1995.

[18]See http://www.io.org/~boo4env/Overview.html/1996.

[19]BVQI, a leading company in registering organizations under the related BS-7750 and the Eco-Management Audit Scheme, asserts that "if a company has ISO 9001 in place, they already have about 70 percent of the implementation know-how of ISO 14001. Many companies will integrate the two standards to have a more complete management system that will cover more of their business activities" [from the BVQI Internet Web site (http://www.bvqi.com)]. In addition to ISO 9000, some sectors might be attracted to merge other standards and processes under the ISO 14000 umbrella. For example, in an industry such as food processing, a reconciliation of quality, health, and environmental systems under ISO 9000/14000 could be attractive, especially for small business where regulatory duplication is particularly onerous.

[20]ISO 14001, 14004, and the audit standards ISO 14010–12 were approved in September 1996. Notwithstanding the early status of the ISO 14000 standards, many countries have already begun the process of deploying certification and registration procedures, and many companies have begun the process of preparing their EMSs for certification.

[21]Such value-added activities could include consulting on technology, loss control, pollution prevention, quality improvement and energy conservation initiatives, training, and many other services that would naturally come to light as part of the process of conducting the EMS audits in a number of companies.

REFERENCES

Allenby, B.R. 1994. Integrating environment and technology: design for environment. Pp. 137–148 in The Greening of Industrial Ecosystems, B.R. Allenby and D.J. Richards, eds. Washington, D.C.: National Academy Press.

Allenby, B.R., and D.J. Richards, eds. 1994. The Greening of Industrial Ecosystems. Washington, D.C.: National Academy Press.

Ayers, R. 1997. The life-cycle of chlorine. Part I. Chlorine production and the chlorine-mercury connection. Journal of Industrial Ecology 1:81–94.

Baram, M.S., P.S. Dillon, and B. Ruffle. 1992. Managing Chemical Risks: Corporate Response to SARA Title III, Rev. Ed. Chelsea, Mich.: Lewis.

Bond, G.G. 1995. Product stewardship shifts into high gear. Chemical Engineering 102:78–84.

Chemical Week. 1991. Product stewardship: exploring the "how-to." December 11, pp. 13–16.

Council of Logistics Management. 1993. Reuse and Recycling—Reverse Logistics Opportunities. Oak Brook, Ill.: Council of Logistics Management.

Dillon, P.S. 1994. Implications of industrial ecology for firms. Pp. 201–207 in The Greening of Industrial Ecosystems, B.R. Allenby and D.J. Richards, eds. Washington, D.C.: National Academy Press.

Fisher, M., W.J. Bell, L.M. Dalberto, A.J. Greenfield, R. Jaikumar, P. Kedia, R. Mack, and P. Prutzman. 1983. Improving the distribution of gases with an on-line computerized routing and scheduling optimizer. Interfaces 13:4–23.

Hart, S. 1997. Beyond greening: strategies for a sustainable world. Harvard Business Review 75:66–76.

Hindle, P., B. De Smet, P.R. White, and J.W. Owens. 1996. Managing the environmental aspects of a business: a framework of available tools. The Geneva Papers on Risk and Insurance 80(July): 341–359.

Jaffe, A.B., S.R. Peterson, P.R. Portney, and R.N. Stavins. 1995. Environmental regulation and the competitiveness of U.S. manufacturing: what does the evidence tell us? Journal of Economic Literature 33:132–163.

Klassen, R.D., and C.P. McLaughlin. 1996. The impact of environmental management on firm performance. Management Science 42:1199–1214.

Kleindorfer, P.R. 1997. Market-based environmental audits and environmental risks: implementing ISO 14000. The Geneva Papers on Risk and Insurance 83(April):194–210.

Kleindorfer, P.R., and E.W. Orts. 1998. Information regulation of environmental risks. Risk Analysis 18(2):155–170.

Kunreuther, H.C., and E.H. Bowman. 1997. A dynamic model of organizational decision making: Chemco revisited six years after Bhopal. Organizational Science 8:404–413.

McNulty, P.J., L.C. Schaller, and K.R. Chinander. 1998. Communicating under section 112(r) of the Clean Air Act Amendments. Risk Analysis 18(2):191–198.

Munasinghe, M.P.C. 1996. Sustainable energy development (SED): issues and policy. In Energy, Environment and the Economy: Asian Perspectives, P.R. Kleindorfer, H.C. Kunreuther, and D.S. Hong, eds. Cheltenham, U.K.: Edward Elgar.

Rosenthal, I., and D.F. Theiler. 1998. Use of an ISO 14000 option in implementing EPA's rule on risk management programs for chemical accidental release prevention. Risk Analysis 18(2): 199–204.

Snir, E.M. 2001. Liability as a catalysis for product stewardship. Production and Operations Management, forthcoming.

Ulrich, K.T., and S.D. Eppinger. 1995. Product Design and Development. New York: McGraw-Hill.

von Hippel, E. 1988. The Sources of Innovation. New York: Oxford University Press.

White, P.H., M. Franke, and P. Hindle. 1995. Integrated solid waste management: a lifecycle inventory. London: Blackie.

Willums, J., and U. Goluke. 1992. From Ideas to Action: Business and Sustainable Development. Oslo, Norway: ICC Publishing and Ad NotamByldendal.

Wu, H., and S.C. Dunn. 1995. Environmentally responsible logistics systems. International Journal of Physical Distribution & Logistics Management 25:20–38.

Simulation Models for Information Sharing and Collaboration

JOSEPH A. HEIM

Globalization and technological innovation have become catch phrases for the collection of competitive challenges, changes, and shocks that our economy is beginning to encounter. For manufacturing, broadened competition and technical advances are succinctly reflected in two metrics: *time*, the profound pressure to respond to the rapid ebb and flow of the market; and *complexity*, the expanded scope and functionality of products and the new bodies of knowledge that influence products and processes.

All of these changes present major challenges to manufacturers, but they are notably difficult for small firms. Here simulation modeling is presented as a cost-effective way for small firms to acquire the knowledge they will need to become competitive manufacturers in the evolving global economy. This paper is based on research the author conducted at Purdue University and the University of Washington (Heim, 1994).

SMALL MANUFACTURING ENTERPRISES

Manufacturing firms are commonly grouped according to the type and volume of goods they produce, the number of people they employ, and their total annual sales. These criteria tend to be positively correlated. For instance, most aircraft and automobile manufacturers would cluster at one end of the spectrum, whereas most of the companies that design and build production machinery for those large firms would be found at, or near, the opposite end of the scale. Accordingly, we would expect the firms supplying components and machinery to generally have lower sales and fewer employees than their customers, and that is indeed the pattern.[1]

Another contrast between small manufacturing enterprises (SMEs) and larger firms is the scale and scope of resources they have to address current problems and pursue new opportunities. Most small firms vest most of their intellectual capital and technical talent in the management of day-to-day operations, keeping the company alive. They have few resources left to develop skills with newer technologies or to thoroughly consider new market opportunities (National Research Council, 1993).

How, then, can we expect small firms to understand and anticipate the performance expected of them when the systems in which they participate—the markets, products, partners, technologies, and processes—are in such flux and so dispersed? How can SMEs *affordably* access the information and knowledge they need to succeed in the global economy? How will they assimilate that new information and make it an integral part of their organizational knowledge? What tools and mechanisms can we provide to facilitate the learning needed to adopt and implement new methods, standards, technologies, and techniques? And how can we efficiently share critical information and knowledge without compromising the intellectual property rights of the organizations that have created the information?

INFORMATION SHARING

At a basic level, there are two kinds of processes for sharing information and knowledge: synchronous processes, in which people are the agents for the instantaneous exchange of knowledge and information, and asynchronous processes, in which an intermediary technology carries an encoded representation of the information and knowledge for retrieval at some time in the future. Common synchronous sharing mechanisms include face-to-face meetings, telephone conversations, and videoconferences. Text-based materials, databases, audio and video recordings, and models of various kinds are some of the ways we share information asynchronously. Obviously, computers and electronic means can be used to facilitate both synchronous and asynchronous exchanges. They can also be used, to some extent, to increase the number of people involved in synchronous exchanges.

If we examine these two information-sharing categories closely, it is evident that the critical difference between them is the immediacy of interaction and, therefore, the types of feedback and dialog each accommodates. Dialogs between people, as well as dialogs with other kinds of information agents (such as Internet search engines), are inherently ambiguous. The key difference is that human-to-human communication relies on various resources to detect and repair communication failures, and these mechanisms are generally lacking when people represent only one side of the interaction pair (Suchman, 1987).

Synchronous processes support high interactivity and debate, but do not provide a good format for extended thought and reflection. On the other hand, asynchronous approaches more readily accommodate protracted investigation and consideration, but do not afford the animated and tightly coupled communications of a synchronous exchange. To share new, complex information, synchronous processes are the preferred method of transfer because they assure that error-free information has been received, and, most importantly, understood in the shortest amount of time. But synchronous processes are also the most expensive, because the ratio between information source and recipients tends to be much lower than with asynchronous methods. Consider, for example, successful telephone conference calls (an example of synchronous communication). If the information to be shared is complex, you can expect more questions (correlated more or less to the number of participants) and you must allocate time for extended explanations. The entire set of questions and responses during the call will be of little value to some because of prior preparation or knowledge of the topic, but all parties must endure the "education" of each member.

For the small manufacturing firm, the objective of sharing information and knowledge is to improve capabilities and performance. In most cases, however, improvements will not happen as a consequence of *access*, but rather as a result of *assimilation*—the new information must become an integral part of a firm's organizational knowledge, and it must be applied effectively. Assimilation of new information or knowledge is accelerated when people are able to actively examine, question, test, and understand its applicability within the context of *their own enterprise* (Papert, 1980). Therefore, the manner in which we package and provide access to information is critical, because the mechanisms we use must help people learn.

For small manufacturers, the attractiveness of any such mechanism will be strongly influenced by its cost, the resources and expertise needed to use it, and the time it takes to obtain worthwhile results with it. An attractive methodology would most likely be a compromise between a highly coupled, dialog-rich learning environment and a more cost-effective, self-directed, and self-paced method of exploration and discovery. For instance, simulation could be used to construct a waste management system that meets the particular needs of a smaller manufacturing firm and conforms to environmental regulatory guidelines.

Flight training, power plant operation, and various other types of simulation models have demonstrated their ability to help people learn by providing a computer environment in which they can experiment without the fear of consequences they would encounter otherwise (De Geus, 1992). But special skills and resources are required to construct simulation models that are applicable to the specific context of a firm. These skills and resources are not found within most small manufacturing firms, and even if a firm has the requisite tools, the cost of developing models can be difficult to justify. One approach might be to have "someone" else build the models and then distribute the models to those with whom we wish to share. In the next section we look at why that may not be effective.

MODELS, MODEL BUILDING, AND LEARNING

Models are typically used in manufacturing to gain a better understanding of the possible interactions and consequences when certain choices are made—that is, direct support for decision making in a computational environment. But many decision makers are not comfortable with models constructed by others; they want to be assured that their ideas and knowledge are represented and indeed reflect the situations for which they are responsible. Furthermore, learning takes place when people discover for themselves the relationships and contradictions between observed behavior and their perceptions of how the world should operate; they benefit from experimentation and testing the scenarios they define. At some level, individuals and organizations must construct their own representations if they are to have confidence in the results obtained (Morecroft, 1992).

But creating comprehensive, illuminating models of complex interacting systems can be expensive. Building and validating models often takes longer than expected and requires special technical skills such as computer programming. And when models are constructed to answer questions of a nonrecurring nature, they are difficult to justify because most of the return on modeling investment must be recouped by the one-time use of the model.

For smaller firms, the adoption of model-based learning and decision making is a kind of Catch-22 situation: They are inexperienced users of sophisticated modeling techniques, so they are unable to justify the type of investments needed to construct and maintain models. The modeling approach we discuss presents one response to this pattern. It is a way that small firms can construct models that provide an environment in which users can learn by experimentation. By playing with their models and conducting their own what-if scenarios, they will acquire a better understanding of their particular world and improve the capabilities of their organization by integrating new information and knowledge (Wegner, 1997).

COMPONENT-BASED MODELING

Much of the cost associated with modeling can be attributed to the craftsmanlike approach we take toward model construction and software development in general. In the early 1800s, classical manufacturing was in a similar situation: Products were made by skilled craftsmen, each component fashioned in a cut-to-fit fashion, so product costs were high. Interchangeable component parts ushered in the Industrial Revolution. Because skilled workmen were no longer required for product assembly, the cost of manufactured products were significantly lower, and, accordingly, many more people were able to afford manufactured products.

Model building (and software in general; see Cox, 1996) is still in the craft mode of production. It needs to move from the classical handiwork approach to an industrial-based method of fabrication from interchangeable component parts.

Model "assemblers" would integrate the model component parts in a plug-and-play manner, thus minimizing the time, cost, and expertise required to construct comprehensive models within the context of their organization. We examine a bit more closely how such a component-based modeling approach might work.

PLUG-AND-PLAY MODELING

The phrase plug and play is often associated with the idea of adding new components to a personal computer and having them interoperate automatically, with no complicated efforts on the part of the user. A somewhat more accessible metaphor to explain plug and play is the common home stereo system.

Although we have to be careful to not push the analogy too far, simulation models of complex systems could be constructed in a manner similar to how we create audio systems for playing and recording music. Industry guidelines dictate standard physical connections and electrical characteristics of stereo components, enabling users to create a wide range of audio systems. System configurations can range from the simple—adequate for listening to broadcast radio, to the complex—able to present rich sound that is difficult to distinguish from a live performance. We can easily add new components to a system, substituting higher-fidelity components where we believe that we gain the most benefit, or replacing multipurpose subsystems (e.g., a combined preamplifier, amplifier, and tuner) with individual components that provide the same functions with greater control and precision (e.g., the ability to adjust frequency curves).

The basic result of each configuration is the same, providing the ability to hear broadcast or recorded music. The difference is in the precision, quality, and fidelity of reproduction, the results of the particular component parts we used. Interoperability of the components is based on well-defined protocols for communication.

Similar standards and protocols are being developed that will allow simulation models to interoperate in much the same way. Recent advances in data communication software, networks, computers, and programming language technologies (National Research Council, 1994) provide an opportunity to develop an open-systems architecture for integrating simulation model component parts. The sections that follow introduce an architecture that could provide a vendor- and language-neutral foundation upon which model builders could construct comprehensive systems models using component models accessible on the Internet.

MODELS AS COMMUNICATING OBJECTS

First of all, where would model assemblers obtain the component parts? We believe that a palette of models, from which enterprise models would be constructed, might come from public institutions such as universities and national labs, or commercial developers and industry-sector groups (e.g., a machine tool builders association). In some circumstances, the model components would be gathered from various sources and assembled on a single computer system. The scope of the components could range from a comprehensive model that reflects most operational aspects of a piece of production equipment, or a mathematical smoothing algorithm that would be one component of a customized forecasting system. In other cases, the proprietary nature of the information would suggest that some component parts would be licensed or used only with permission. For instance, a manufacturing company considering the purchase of new production equipment could examine the consequences of using that equipment in its plant by linking machine tool models provided by various vendors with the company's current production system model (assembled from component parts obtained in the market). The company could then quickly determine the overall impact on performance (a systems perspective) as well as consider interactions between the new equipment under consideration and machinery it already owns.

Our work on simulation model component parts has three fundamental concepts: (1) models are *objects*, (2) models communicate with one another in *client–server* relationships by passing messages, and (3) each model is represented by an *agent* that explains the capabilities of the model and assists with integration of that model. A few words about objects, client–server relationships, and agents will help clarify this approach to component model integration.

Objects

Three basic principles define object-oriented programming: *encapsulation*, the way objects hide implementation and associated data but advertise functionality; *message passing*, a strict protocol by which objects communicate and request performance of advertised functionality by other objects; and *classes and inheritance*, a means of organizing the kinds of objects that are created to maximize code reuse and minimize maintenance efforts (Goldberg and Robson, 1983). A significant benefit of object-oriented programming is the reduction in *cognitive distance* between the world that one wants to represent in the computer and the mechanisms that are available to accomplish that representation. Object-oriented programming does this by preserving the decomposition of the system in the computer code that is created.

For instance, for creating a model of a manufacturing plant, one might want to represent the machines, routings, work in process, the tools and fixtures, the customer orders, and the workers. Using an object-oriented approach, we identify the "things" or entities in the system and the relationships among the various entities—what they do to accomplish the objectives of the system. Our application entities, or *objects*, might be the customer orders, workers, routings, machines, tools, and tooling fixtures, and work in process. Each of these objects would be represented by a coherent chunk of code that contained both the *functionality* for that object and the *state* of the object. The functionality would be that set of activities that the object would perform if asked, and the state of the object would be the value of all variables describing that particular object. For example, the customer order object would be able to answer questions about its internal state such as "what kind of product are you?" or "what is your due date?" The machine objects might respond to messages such as "begin busy state with this order." The result of that message would be that the machine object receiving the message would change the value of the internal-state variable representing its busy or idle status.

Unlike procedural approaches to program development, in object-oriented programming the state and implementation of the functionality are hidden. The only way for one piece of code (i.e., object) to change the state of another object is by sending a message requesting that change. The focus is on the objects in the system and the activities they must accomplish. Object-oriented programming languages provide constructs that allow programmers to maintain the relationships between chunks of code and things in the world to be represented. A major milestone of the research is developing the ability to encapsulate individual models so that they might function as objects and exchange messages to accomplish modeling tasks without the user of the model needing to address details of integration and implementation.

Client-Server Relationships

The most widely dispersed example of client–server relationships is the World Wide Web. *Browser* software packages, such as Netscape Navigator and Microsoft Internet Explorer, are the clients, and the applications providing information and data are the servers. We say that the *requests* come from *clients* and that the *server responds* to those requests. In more advanced applications the client or server attribution is likely to be dynamic, based on the context of the communicating program processes: Sometimes a program process will be a server, but in other contexts it may also be a client of another program process.

For instance, it is easy to envision a situation in which our work-in-progress (WIP) order objects, machine objects, and material handling objects could be both clients and servers. An order (client) requests that a machine object (server) perform some transformation; the machine object (client) in turn requests that the material handling system (server) transport the WIP order from its present location. In all cases, servers are not concerned with the source of the request (in object terms, the request is a message) except to know where the results of the request must be returned. Accordingly, the client is not concerned about the manner in which its request is accomplished by the server (the server encapsulates, or hides the manner in which it computationally achieves its activities). This kind of relationship among program processes provides great flexibility for implementation. Because clients have no concern about internal changes to implementation, revisions and improvements to the server side of the relationship can proceed independently. The server is only responsible for continuing to respond to its previously advertised capabilities (services) in the agreed-upon manner (the protocols for exchange).

This means that we can substitute modeling implementations, even going so far as to move the model to a new platform for higher-speed computation, and the users, the remote clients of that model, will not have to make changes to the manner in which the models interact. Improvements and maintenance may proceed independently of users of the model services.

Model-Building Agents

The purpose of model agents is to facilitate the assembly of more complex models. Model agents know what their model object can accomplish, what data they need to perform those actions, and what information the model will provide as it executes. For example, to construct a network model,[2] the model agents representing all of the necessary model objects are downloaded. These agents could be commercial products developed and maintained by a for-profit firm, or they could be developed by national labs or university research programs. Obviously, a certain amount of infrastructure would have to be developed to support the creation and distribution of the model objects. If we were to construct a network model of a workcell, some of the object models needed would be models for each of the machines in the cell, a control object to manage the activities of the cell, an order object to schedule release of parts to the cell, and a quality assurance model to reflect cell performance. The agents will configure the interface of their respective model objects and provide the information necessary to configure the network model. The agents also help the user select the appropriate model objects from those available on the network by providing semantic information about the model objects they manage. The builder of the individual models creates each model agent using tools and templates also developed in this project.

The agent is created and maintained separately from the model. For instance, there could be several implementations of the same modeling function and all could be represented by the same agent. The agent would help the user select the most appropriate implementation based on execution speed, size of the task to model, and ancillary capabilities (e.g., graphical or animation output).

As the specificity, functionality, and intellectual property content of models from equipment suppliers and other commercial sources grows, the importance of retaining control and restricting access will become important. The information and intellectual property captured in models would be of significant interest and value to competitors. Agents can provide the intellectual property controls and accounting mechanisms needed to allow customers considering adoption of their equipment access to their high-fidelity models. Vendors could charge users for access to their models and rebate those costs if the equipment modeled was subsequently purchased from them.

Our basis for integrating simulation model component parts then is to create the framework and methodology in which individual models can become message-passing objects that communicate with one another as both clients and servers. Each model has associated with it an agent that describes the capabilities of the model, its constraints, and data needs as well as the data it produces and any coordination requirements. The agents for the models also generate the interface programming logic needed to participate in the distributed modeling activity.

SUMMARY

For manufacturing firms, the consequences of competition and technological innovation are reflected in the profound pressure to respond synchronously to the rapid ebb and flow of the market, the expanded scope and functionality of products, and the distributed nature of the entire product realization process. Decisions become less intuitive as the complexity of the systems increase, the time to make good decisions is shortened, and, for small firms, the cost for incorrect decisions can be life threatening. New information must be assimilated and the organizational knowledge, skills, and expertise must quickly adjust to use that information to advantage.

Models have always been used to reduce the time, cost, and risks associated with decision making, but they can also be an effective mechanism for formally transferring new information. In this paper we have examined object-oriented component parts as a mechanism for constructing comprehensive simulation models that reflect the context of the individual manufacturer. We believe that our approach has particular relevance for small firms that must quickly acquire appropriate information and integrate that information with the unique knowledge, talents, and skills they currently possess while minimizing investment in additional resources.

NOTES

[1]It is all too easy to ignore the individual influence of the smaller manufacturing firm, but their aggregate impact is readily apparent. In the United States there are more than 375,000 manufacturing establishments providing employment for approximately 40 percent of our manufacturing work force. Of these 375,000 enterprises, 99 percent have fewer than 500 employees; 92 percent employ less than 100; and 50 percent employ 10 or fewer employees. Furthermore, the trend from 1980 to 1990 has been toward smaller rather than larger firms: The number of companies with less than 20 employees has grown more than 26 percent while those with more than 500 employees has declined 16 percent (National Research Council, 1993).

[2]A network model is the name given to the collection of model objects that are linked to one another via the Internet.

REFERENCES

Cox, B. 1996. Superdistribution: Objects as Property on the Electronic Frontier. Reading, Mass.: Addison-Wesley.

De Geus, A.P. 1992. Modelling to predict or to learn? European Journal of Operational Research 59(1):1–5.

Goldberg, A., and D. Robson. 1983. Smalltalk-80: The Language and Its Implementation. Reading, Mass.: Addison-Wesley.

Heim, J.A. 1994. Integrating distributed models: the architecture of ENVISION. International Journal of Computer Integrated Manufacturing 7(1):47–60.

Morecroft, J.D.W. 1992. Executive knowledge, models and learning. European Journal of Operational Research 59:9–27.

National Research Council. 1993. Learning to Change: Opportunities to Improve the Performance of Smaller Manufacturers. Washington, D.C.: National Academy Press.

National Research Council. 1994. The open data network: achieving the vision of an integrated national information infrastructure. Pp. 43–111 in Realizing the Information Future: The Internet and Beyond. Washington, D.C.: National Academy Press.

Papert, S. 1980. Mindstorms. New York: Basic Books.

Suchman, L.A. 1987. Plans and Situated Actions. The Problem of Human-Machine Communication. Cambridge, U.K.: Cambridge University Press.

Wegner, P. 1997. Why interaction is more powerful than algorithms. Communications of the ACM 40(5):80–91.

Opportunities for Collaboration
and New Technologies

Industrial Research and Development Collaborations
Increasing Environmental Knowledge for Competitive Advantage

PAUL C. KILLGOAR, JR.

Most private-sector collaborations are driven by the motive to gain information and knowledge that can be integrated into a company's operations and used to competitive advantage. In many ways collaborations may seem to be paradoxical, because collaborations also require sharing information. It is that paradox that makes collaborations such a challenge. The history of research and development (R&D) collaborative efforts in the automotive industry illustrates how some of the challenges inherent in collaborations were met, where the potential for the use of information technology in collaborative effort lies, and how these efforts have contributed to improving environmental performance.

HISTORY OF COLLABORATION IN THE AUTOMOTIVE SECTOR

During the 1960s, when Chrysler, Ford, and General Motors (GM) dominated the world market in trucks and cars, collaboration among the Big Three was unimaginable. Competition among the Big Three was strong, there were no external threats to U.S. automakers, and there was no reason or incentive to collaborate. Even if there had been a reason, antitrust law would have prevented collaborations because they were seen as undermining competitiveness. In fact, employees were regularly reminded of their obligations and responsibilities under the antitrust laws, and lawyers often were present during trade-group meetings to answer questions and provide advice on matters related to antitrust concerns.

In the 1970s the situation changed. The Big Three started to lose their world leadership. Japanese automakers began to expand their export markets and establish their automobiles as quality products. In addition, regulation began to affect

U.S. industry, particularly requirements imposed by the Clean Air Act and the Corporate Average Fuel Economy program. Chrysler, Ford, and GM were all challenged to develop exhaust treatment technologies and increase the fuel efficiency of their fleets. The costs required to develop these technologies were significant and independent of company size; thus, larger manufacturers, such as GM, were able to spread these costs over large sales volumes, and therefore had some competitive advantage. However, even though foreign competitors were required to meet the same U.S. regulations, they could legally collaborate in their home countries and often were encouraged to do so by their governments. This ability to collaborate gave foreign companies (often smaller in size) a significant competitive advantage, not only in the automotive industry but in other U.S. industries as well.

To address these competitive concerns, the U.S. government passed the Cooperative Research and Development Act in 1984. This act allowed U.S. industries to undertake collaborative research in noncompetitive or precompetitive areas. Under this act, Big Three management got together and set the stage for various collaborations. Engineers in all three companies worked out the details, identifying the key noncompetitive areas for collaboration. From this effort, two areas emerged with potential for collaborations: The first area related to technologies where there was no customer differentiation, and the second area related to R&D directed at societal good, such as environmental improvements and occupant protection.

The cigarette lighter is illustrative of the first area of collaboration. At one point, the Big Three had about 20 different cigarette lighter designs in use, although the sale of a car probably never hinged on the basis of a cigarette lighter. It made sense, then, for the Big Three to come to an agreement on what the functional and design characteristics of a cigarette lighter should be. From this work it became possible to reach agreement on two or three common designs for cigarette lighters. Decisions such as this helped to provide an economy of scale and improved quality for certain automotive components.

The second area of collaboration, and environmental concerns, in particular, have driven the majority of the automotive industry's joint efforts. One early example of this type of collaboration was the formation of the Auto/Oil Air Quality Improvement Research Program to study the effects of fuel composition on vehicle emissions. This work was used by the automobile and oil industries as a basis for discussion with the federal and California state governments in setting standards relating to fuel.

Figure 1 shows the history of collaborations in the automobile industry beginning with the enactment of the 1984 Cooperative Research and Development Act. The 1986 Technology Transfer Act expanded collaboration between industry sectors and led to the creation of the Auto Steel Partnership and the Automotive Composites Consortium. These early efforts laid the groundwork for the many R&D collaborations to follow. Opportunities to expand the collaborations

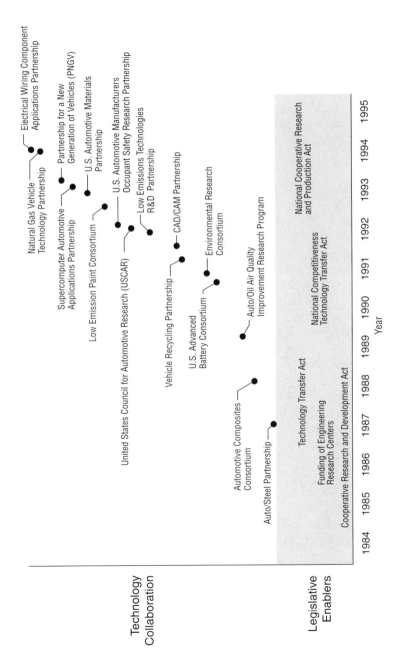

FIGURE 1 Chronology of auto industry collaborative research. SOURCE: USCAR, 1997.

to include national laboratories were facilitated by the 1989 National Competitiveness Technology Transfer Act.

Several automotive consortia and partnerships were formed after the 1989 act, all of which related to environmental concerns. For example, the Environmental Research Consortium focuses on measuring emissions from vehicles and assessing soil remediation technologies. The U.S. Automotive Materials Partnership is looking at material alternatives that lighten the weight of vehicles. The Low Emission Paint Consortium is working on developing solvent-free clearcoat technology, and the Vehicle Recycling Partnership is establishing procedures to increase the recycling of end-of-life vehicles.

PRACTICAL CHALLENGES IN COLLABORATIONS

The process of establishing each of these collaborations was difficult. All posed administrative, definitional, and legal challenges. The administrative challenges ranged from scheduling meetings and getting contracts signed to handling intellectual property concerns. To address these issues and to provide focus for the precompetitive research of the Big Three, the United States Council for Automotive Research (USCAR) was established in 1992. With this umbrella organization in place to deal with administrative issues, the researchers were freed up to focus on their work.

As it turns out, the administrative challenges were among the more transparent ones, and a number of them were identified in early collaborations as factors that could stall progress.

The first issue was the difficulty of establishing trust. There is a common belief in collaborations that each participant is trying to extract as much information as possible from the other collaborators while sharing the minimum of their own knowledge. Building trust takes time and lots of face-to-face meetings.

The second issue was the difficulty of defining the bounds of the research. Because the collaborative projects involved precompetitive technologies, the participants had to understand issues such as what portion of the research was precompetitive, where the research crossed into competitive areas, and how relevant background information could be identified and shared.

The third issue was related to intellectual property ownership and how such property could be used by non-U.S. subsidiaries. There were major legal issues to hammer out, particularly in the earlier consortia, but they are less contentious today because of the long history of collaborative activities in the automotive industry.

The final issue was the complex nature of technical project management. Collaborators had to define the scope of work to be undertaken, set schedules for milestone completion, and assign responsibilities for the project tasks. Much of this work involved sharing background information, where, as pointed out above, trust was needed for progress to occur.

Some of the issues discussed above became even more difficult when the collaborations extended beyond the Big Three. For example, USCAR is a partnership of the Big Three, and although suppliers are not part of the organization, they can participate on individual projects as technical teams deem necessary. The most significant complication of extending the circle of participants was in obtaining suitable legal agreements covering the ownership of intellectual property. In comparison, the involvement of governmental entities and universities is easier to finesse, and occurs more frequently. For example, when one of the consortium projects wanted to acquire data on real-world vehicle emissions and automobile ownership, it brought the Michigan State Police and Department of Motor Vehicles into the consortia. These organizations were interested in the data; and in exchange for the results, they provided logistical and data support for the project.

Under the Clinton administration the health of technology was raised as a critical national priority, and in 1993 the automobile industry saw the creation of the Partnership for a New Generation of Vehicles (PNGV). The PNGV is an ambitious collaboration involving the U.S. government and its laboratories, USCAR, suppliers, and universities in an effort to develop next-generation cars that are fuel efficient and environmentally friendly.

The long-term research goal of the PNGV is to develop vehicles that meet consumer needs for safety, quality, performance, utility, and affordability and that achieve up to three times the fuel efficiency of today's comparable vehicles (identified as the Ford Taurus, Chrysler Concorde, and Chevrolet Lumina). The more immediate goals of the PNGV are related to manufacturing and the implementation of near-term advances. In manufacturing, the goal is to pursue advances that can reduce production costs and development lead times for new cars and trucks. The near-term goal is to pursue technological advances that can lead to improvements in the fuel efficiency and emissions of conventional vehicles. Both of these goals are aimed at cars being built today, and in essence, they call for implementing new technologies as they are developed.

The implementation challenge of the PNGV is the responsibility of each party involved. The question for each is how to get the information from the collaboration back to their organization where it can be implemented and used to gain competitive advantage. The concept sounds simple, but the execution can be difficult. At issue is the communication between the groups assigned to the task of developing the technology (the research community) and those responsible for implementing the concepts (the design and manufacturing engineers). For the implementation to work, the researchers need to engage their company's design and manufacturing personnel early on in the process so that a viable implementation plan can be developed. When one considers the vast networks of suppliers who need to be involved in this process, the size and scale of such an implementation can be daunting. However, the process can be aided by the appropriate use of information technology.

INFORMATION REQUIREMENTS IN COLLABORATIONS

Each collaborative program has unique requirements for managing and transferring information, based on the program's goals and participants. Existing consortia provide some insights into differing needs for collecting and sharing information in collaborations. In the Low Emissions Technologies R&D Partnership, the mission is to coordinate R&D efforts on emission control technologies through the exchange of technical information. The overall goal is to identify and develop enabling emissions technologies, drawing from the existing knowledge base and R&D work of the Big Three. In this collaboration the participants communicate via face-to-face meetings, paper documents, and e-mail. E-mail is particularly valuable in some situations because it enables all participants to receive the same information at essentially the same time. Most of the information comes from the results of collaborative research being conducted by the Big Three, and each participant is responsible for archiving the data.

In the Low Emission Paint Consortium, the focus is on paint-related technologies that can reduce or eliminate solvent emissions. The group has an explicit goal of building a common industry database for the information gathered. The consortium's first collaborative project explores the materials and process issues related to using powder clear-coat paint systems in automobile manufacture. Powder paint systems are solvent free and thus reduce plant emissions, and the consortium has built a highly instrumented pilot facility to test their use. To ensure that all of the necessary expertise is available, the consortium includes manufacturers of paint and paint spray equipment. All participants are involved in developing the materials, processes, and monitoring systems needed to show the environmental impact of specific paints. The ultimate goal is to place proven materials and processes into automotive assembly plants so that plant emissions can be reduced.

The Casting Emissions Reduction Program offers another example of how information is shared in collaborations. The Big Three are dependent on some of the 3,100 foundries in the United States that supply the castings for engines, transmissions, and other parts. The foundry industry is essential not only to the automakers, but to other groups as well, including the Department of Defense (DOD). Emissions from foundries are cause for concern, but no individual foundry is in a position to conduct the magnitude of research needed to examine the potential problems. To address these issues and help protect the national supply base, a collaboration was developed between USCAR and the DOD that included the U.S. Environmental Protection Agency (EPA), the California EPA Air Resources Board, and the American Foundrymen's Society. The consortium has built a state-of-the-art facility at McClellan Air Force Base to look at new environmentally friendly and efficient casting processes designed to meet new regulations expected in the year 2000. The program's goals are to collect and analyze data on casting quality, foundry processing, and foundry emissions and to find processes and technologies that will allow the foundries to stay in business. The information technology challenge in this program is to acquire and organize the data and disseminate the findings to the foundries. The consortium hopes that the involvement of the American Foundrymen's Society will help to speed that process, as it provides a central source for industry information through its publications and meetings.

USE OF INFORMATION TECHNOLOGIES IN COLLABORATIONS

Various information technologies are being used to manage the data derived from the various automotive R&D collaborations. These technologies include those that are used routinely in corporations to communicate, store, and relate information, such as e-mail, CD-ROMs, intranets, databases, and others (see Carberry, this volume). Although information tools are useful in many situations, they may be less so in the early stages of a collaborative project, when it is important to establish a sense of rapport and trust between the participants. Although teleconferences and videoconferences may approach the immediacy of direct meetings, there are nuances in the communication process that are missed when using these devices. Once rapport and trust are established, information technologies may become more valuable in facilitating a collaboration, but in the early stages participants need to meet face to face to make the collaboration work.

A peek into the near future suggests that information-sharing technologies such as groupware show promise for use in collaborations, but until these technologies become more commonly used (not just available) in individual organizations for collaborative purposes, their utility is limited. In addition, such efforts are likely to confront issues related to the diversity of available software packages. So, at the moment, the key to a successful collaboration remains the people involved.

When it comes to archiving and disseminating information, the Internet and intranets are evolving as effective tools. This may be the place where information technology will have its greatest impact—that is, in preserving knowledge in an easily accessible form. However, at the present time, most data, particularly R&D and more technical information, are still exchanged on diskettes, in written documents, and through oral presentations. The challenge in using any method continues to be the need for more knowledge to move out of the research departments and into the larger corporations. Even this, so far, has best been accomplished in face-to-face meetings.

SUMMARY

The majority of the Big Three collaborations have been directed at solving environmental problems. Early collaborative efforts included the manufacturers and government research entities. Today, collaborative efforts include a range of other players—particularly suppliers. Historically, information exchange in these collaborations has been accomplished via face-to face meetings, traditional paper documents, and familiar computer media such as diskettes and databases. More recently, technologies such as the Internet and intranets have shown promise as vehicles for information exchange. However, the promise of these and other new tools for cross-company collaborations will be realized only as personnel in the companies become more familiar and comfortable with them.

REFERENCE

Killgoar, P. 1997. Chronology of Auto Industry Collaborative Research. Paper presented at the National Academy of Engineering Industrial Ecology Workshop, Woods Hole, Mass., July 20–22.

InfoSleuth

An Emerging Technology for Sharing Distributed Environmental Information

GREG PITTS and JERRY FOWLER

The rapid growth of the World Wide Web (or the Web) has profoundly affected the culture of information technology. The enormous increase in the number of users and the proliferation of Web sites have led information technologists to rethink the way that information is delivered via a computer. Previously, a computer mainframe provided data from a single structured database to a small group of users over dedicated lines. Today, anyone with a personal computer and access to the Web can read information distributed over computer networks, and, more importantly, they also can create and publish new information resources.

This rapid transformation of the culture along with increased expectations of Web users in all disciplines have led to two challenges. First, information discovery and retrieval is a problem that is addressed only superficially by the development and use of Web search engines (such as Alta Vista and Lycos) to handle the huge amounts of data accessible on the Web. Second, the structured databases that were carefully crafted with a small, knowledgeable user community in mind are now potentially accessible and useful to many additional users. These users, however, are unfamiliar with database query languages. The result is a need for information systems that can uniformly deliver both structured and unstructured data, access different structured databases using a single query language and logical structure, manipulate the data from these distributed sources, and communicate with users via software standards and behavior metaphors that they find comfortable.

Environmental information systems are no exception to this cultural revolution. Currently, it is very difficult to share environmental data because the information typically resides on geographically disparate and heterogeneous systems. Because they were designed with little expectation of interoperability or widespread access, these systems are not easy to access by secondary users. This can frustrate attempts to fuse data from multiple sources in the interest of arriving at a comprehensive understanding of environmental conditions and actions.

The Environmental Data Exchange Network (EDEN) project was undertaken to address these issues. A collaborative effort of the U.S. Department of Defense (DOD), the U.S. Department of Energy (DOE), the U.S. Environmental Protection Agency (EPA), and the National Institute of Standards and Technology (NIST), the EDEN project seeks to provide a flexible dynamic system for accessing environmental data stored in diverse distributed databases. The underlying technology for EDEN is an emerging information technology called InfoSleuth™[1] developed at the Microelectronics and Computer Technology Corporation (MCC). This chapter provides background on the InfoSleuth technology and its application in the EDEN project.

INFOSLEUTH

In the past, database research has been focused on the relatively static environments of centralized and distributed-enterprise databases. In these environments, information is managed centrally and data structures are fixed. Typically, the integration of concepts to specific sets of data is well known at the time that a database schema is defined, and data access can be optimized using precomputed approaches. Although these federated database systems support distribution of the resources across a network, they do not depart from the centralized model of a static database schema.

The Web presents a different paradigm. On the Web, there is a tremendous amount of textual information, spread over a vast geographic area. There is no centralized information management because anyone can publish information on the Web, in any form. Thus, there is minimal structure to the data. What structure there is may bear little relationship to the semantics. Therefore, there can be no static mapping of concepts to structured data sets, and querying is reduced to the use of search engines that locate relevant information based on full-text indices.

The InfoSleuth Project at MCC broadened the focus of database research to produce a model that seeks to combine the semantic benefits of structured database systems with the ease of publication and access of the Web (Bayardo et al., 1997). This change in fundamental requirements dictates a pragmatic approach to merging existing research in database technology with research from other computer disciplines. The result is an architecture that operates on heterogeneous information sources in an open, dynamic environment.

Information requests to InfoSleuth are specified independently of the structure, location, or even existence of the requested information. A key to the success of this approach is the development for each application of a unifying ontology for the application domain. This enables the user to bridge the gap between different notions of data and different schema of databases. InfoSleuth accepts requests specified at a high semantic level in terms of the global ontology, and flexibly matches them to the information resources that are available and relevant at the time the request is processed.

InfoSleuth is an extension of previous MCC work, the Carnot project, which was successful in integrating heterogeneous information resources in a static environment (Huhns et al., 1992). In this previous work, MCC developed semantic modeling techniques that enabled the integration of static information resources and pioneered the use of intelligent agents to provide interoperation among autonomous systems. The InfoSleuth Project extended these capabilities into dynamically changing environments, where the identities of the resources to be used may be unknown at the time the system is designed. InfoSleuth observes the autonomy of its resources and does not depend on their presence. Information-gathering tasks, therefore, are defined generically, and their results are sensitive to the availability of resources. Consequently, InfoSleuth provides flexible, extensible ways of locating information while executing a task and deals with incomplete information.

InfoSleuth Technologies

To achieve the necessary flexibility and openness, InfoSleuth integrates the following technological developments:

- **Agent technology.** Specialized agents that represent the users, the information resources, and the system itself cooperate to address the users' information processing requirements, allowing for easy dynamic reconfiguration of system capabilities. For instance, adding a new information source involves merely adding a new agent and advertising its capabilities. The use of agent technology permits a high degree of decentralization of capabilities, which is the key to system scalability and extensibility (Nodine and Unruh, 1997).
- **Domain models (ontologies).** Ontologies give a concise, uniform, and declarative description of semantic information, independent of the underlying representation of the conceptual models of information bases. Domain models widen the accessibility of information by allowing multiple ontologies belonging to diverse user groups.
- **Information brokering.** Broker agents match information needs, specified in terms of some ontology, with currently available resources. Retrieval and update requests then can be properly routed to the relevant resources (Nodine and Bohrer, 1997).
- **Internet computing.** Java programs and applets are used extensively to provide users and administrators with system-independent user interfaces

and to enable ubiquitous agents that can be deployed at any source of information regardless of its location or platform.

Agents

The InfoSleuth system employs a number of intelligent software agents to perform its tasks, concealed from the user by a dedicated user agent. These agents operate independently in a distributed fashion and may be located anywhere over the network in the form of a Java program. Each agent provides a critical capability in the overall system, as described in the following:

- **User agent.** Constitutes the user's intelligent gateway into InfoSleuth. It uses knowledge of the systems' common domain models (ontologies) to assist the user in formulating queries and in displaying their results.
- **Ontology agent.** Provides an overall knowledge of ontologies and answers queries about ontologies. This permits users to explore the terminology of the domain and learn to phrase their queries to obtain useful results.
- **Broker agent.** Receives and stores advertisements of capabilities from all InfoSleuth agents. Based on this information, it responds to queries from agents as to where to route their specific requests.
- **Resource agent.** Provides a mapping from the global ontology to the database schema and language native to its resource and executes the requests specific to that resource, including subscription queries and notification requests. Resource agents exist not only for structured databases, but also for unstructured data sources that serve text or images.
- **Task execution agent.** Coordinates the execution of high-level information-gathering subtasks necessary to fulfill queries and other information management tasks, such as the control of workflow processes.
- **Multiresource query agent.** Uses information supplied by the broker agent to identify the resources likely to have the requested information, decomposes the query into pieces appropriate to individual resource agents, delivers these subqueries to the resource agents, and then retrieves and reassembles the results.

How InfoSleuth Works

Comparing the behavior of InfoSleuth to the administration of a library provides a good analogy for explaining how InfoSleuth works. At the physical level, the infrastructure of a library is its bricks, mortar, windows, wiring, plumbing, and shelves. Add a few books and you have an information system that is usable. However, a more usable system will have books classified and arranged by their classification.

To take advantage of the information in the library, a librarian needs to be provided with a set of subject headings, such as those provided by the Dewey decimal system, the Library of Congress system, or the National Library of Medicine medical subject headings, and a mapping from the subject headings to the library's holdings. In addition, the books need to be organized according to these subjects, and the librarian needs to be given a resource locator list that identifies where books categorized by a particular subject can be found.

For library patrons to take advantage of the library, they need (in addition to the items mentioned above) a competent library staff person who can assist them in their searches, or at least direct them to the tools required to perform a search.

The InfoSleuth model of dynamic distributed information management has several strong parallels with a library system. A library's physical infrastructure is its building; similarly, a distributed information system comprises the computers and the networks connecting them. The library's books correspond to the databases and other information repositories that InfoSleuth serves. The library's subject headings have an analogy in the domain ontology (stored in the ontology agent's knowledge base) that InfoSleuth agents use to query their resources. Just as libraries benefit from the use of standard classification schemes such as that of the Library of Congress, InfoSleuth benefits from the use of global ontologies. This makes it possible to query diverse databases through a common query model. The broker agent's knowledge base of agent advertisements can be thought of as combining knowledge of the library's card catalog and holdings locator list with an administrator's knowledge of the capabilities of the library staff.

Query Initiation

An InfoSleuth query can be described in terms of a visit to a library. For example, a professor at an academic institution instructs a graduate student to do research on phytoremediation of petroleum hydrocarbons (the specific domain of interest) at the library (i.e., where the Web browser or Viewer application points to the appropriate location to access the InfoSleuth system). The student (the user agent) walks in the front door and asks for help at the information desk. There, a librarian familiar with the collection and the staff (the broker agent) is available to direct the student to a research librarian familiar with the domain (see Figure 1).

Query Processing

The student is told to go either to the engineering librarian, who is in another building, or to the earth sciences section of this library, which may have less information but is not so far away. The student finds a knowledgeable research librarian (execution agent) who helps to refine the query with the help of the subject headings (ontology), distilling the query (multiresource query agent) into specific queries about books and journals. The research librarian then assists the

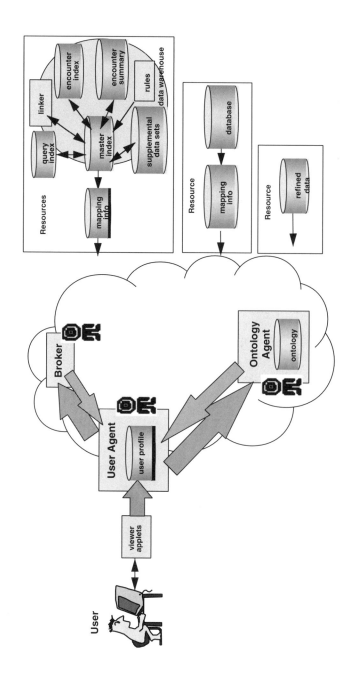

FIGURE 1 A query initiated by the user invokes the appropriate ontology agent and contacts the broker agent.

student by issuing requests to library assistants (resource agents) to retrieve the books. For books that cannot be located, the librarian provides further assistance by issuing an interlibrary loan request (another resource agent) (see Figure 2).

Query Response

When the books are retrieved, they are collected and given to the student (user agent), who then compiles and assesses the information. Eventually, the student provides an analysis in the form of a technical review that collates the results of the search and summarizes them into a single document (multiresource query agent) (see Figure 3). In reviewing the summary, however, the professor questions the validity of a conclusion drawn by one of the authors and requests that the student get the journal or book in question for a more detailed study. The student gets the journal or book in question, and the research query then is completed.

ENVIRONMENTAL DATA EXCHANGE NETWORK PILOT PROJECT

System Characteristics

The broad collaborative effort of the EDEN Project focuses on two areas. The first area is to describe the content of environmental information. What is it? What does it mean? What is its quality and utility? Why was it created and how? What specific data are desired from the data sources? The second area is to develop a means for sharing this information without incurring the financial and technical burdens of redesigning database systems or maintaining redundant databases. This is the immediate focus of EDEN's pilot project.

To this end, EDEN Project participants have chosen the area of hazardous waste remediation for a pilot demonstration. Each government agency is contributing one or more databases (see Box 1). Portions of these databases will be made accessible through InfoSleuth.

Using intelligent InfoSleuth software agents and Java applets to access and retrieve information from disparate data sources, the EDEN pilot system supports a dynamic environment in which databases can be added or removed without affecting the basic behavior of the system. Thus, the project can be developed from a small initial group of databases and allows for additional databases to be added. The resulting technology demonstration is of immediate use in providing access to distributed environmental data resources via the Internet, as well as in guiding future expansion of the information system.

In addition, InfoSleuth provides a common vocabulary and a general query ability that simplifies the exchange and sharing of information among organizations. By establishing the common vocabulary, widely differing information resources can be "mapped" and easily accessed by a sophisticated system of

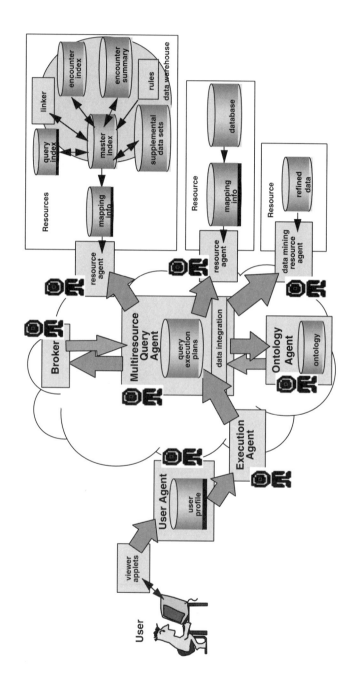

FIGURE 2 The query is interpreted and the information requested of the appropriate resources.

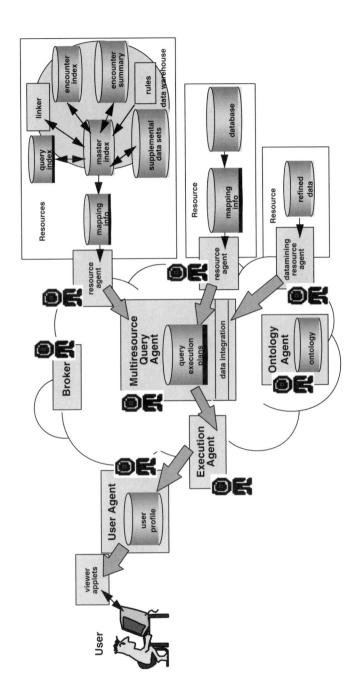

FIGURE 3 The results are returned and compiled for the user.

BOX 1
EDEN Pilot Databases

1. The EPA's Coordinated Emergency Response Cleanup and Liability Information System provides summary data concerning site characterization and contamination of sites on the Superfund National Priorities List (NPL).
2. The EPA's Innovative Treatment Technologies contains information relating to technologies for remediation of chemical contaminants, chiefly emphasizing sites on the NPL.
3. The EPA's Hazards Data relates chemicals to their health effects on humans.
4. The DOD's Installation Restoration Data Management Information System provides site-specific data on contaminated sites at military installations managed by the U.S. Army.
5. The DOD's Environmental Resource Program Information Management System provides site-specific data on contaminated sites at military installations managed by the U.S. Air Force.
6. The DOE's Environmental Remediation Information System provides data on contamination at Idaho National Environmental and Engineering Laboratory.
7. The DOE's Oak Ridge Environmental Information System provides data on contamination at Oak Ridge National Laboratory.
8. The European Environmental Agency's (EEA) Basel Convention database relates data describing transboundary shipment of hazardous waste in accordance with the Basel Convention on that subject.

The pilot will also leverage work being done in two other important environmental information projects that are currently under way:

- The EPA is developing an Environmental Data Registry (EDR) for describing data elements. The EDR serves as an important foundation for addressing the value mapping problem between data sources and the global ontology.
- The General European Multilingual Environmental Thesaurus, under development by the EEA, provides structured knowledge that makes possible translation of queries between European languages.

software agents that advertise, broker, and exchange the data requested by the user. In this way, the EDEN pilot system will provide uniform access to existing information resources without imposing requirements for restructuring or incurring the significant cost of conventional database integration.

Development Plan

The development plan for the EDEN pilot involves the following steps:

1. Create a conceptual model that will become the application's domain ontology. To the extent possible, the ontology is being constructed using terms from the General European Multilingual Environmental Thesaurus

that are applicable to the domain of waste remediation. A graphic depiction of the ontological model that supports a set of queries chosen to show off the capabilities of the pilot is found in Figure 4. This figure shows the primary entities of the ontology linked by a collection of relationships that provide the best abstraction of the data to be found in the databases chosen for the pilot.

2. Develop mappings between each of the identified databases and the domain ontology and then configure a set of resource agents, each of which uses the appropriate mapping to translate between its database resource and the common vocabulary provided by the domain ontology.

3. Take advantage of the Environmental Data Registry (EDR) to assist in resolving value mapping between the different ways that database designers have used for storing data values that express the same concept (such as conversion between English and metric measures or reconciling different ways of identifying chemical contaminants).

4. Develop a flexible yet simple query tool that allows a user to pose queries over the domain ontology and retrieve answers whose appearance may be customized for an individual or group. The user interface is constructed using Java, which makes it portable across numerous operating systems

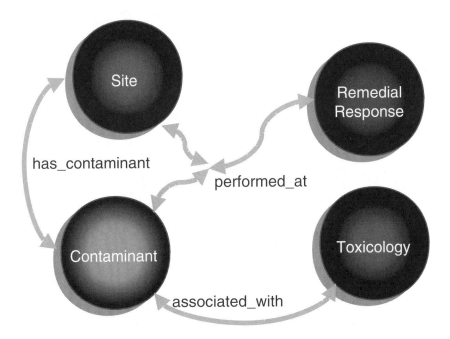

FIGURE 4 The ontological model that supports a set of queries for the selected databases.

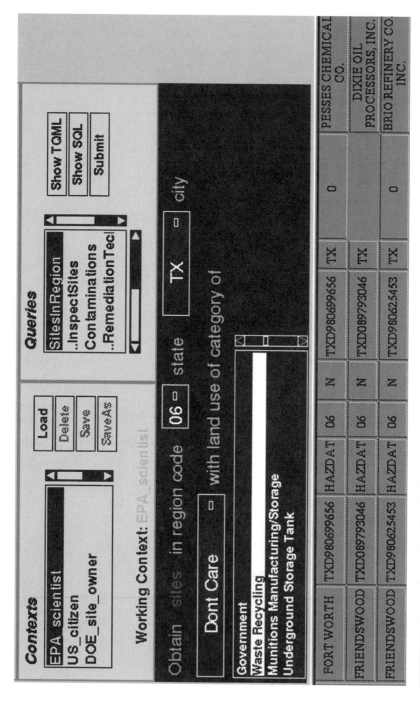

FIGURE 5 A sample user interface for identifying remediation technologies.

and graphic user interface environments. By using Java, it is possible for an end user to issue queries to the system with no more hardware or software than is necessary to support a Java-capable Web browser. Figure 5 depicts a user interface that has been configured to support a particular sequence of queries relating to the identification of remediation technologies associated with particular sites. The displayed results are a fragment of those retrieved from a demonstration system that accessed the Coordinated Emergency Response Cleanup and Liability Information System, the Hazards Data, and Innovative Treatment Technologies.

Two other complementary efforts are of importance. These are investigating the utility of integration of data analysis capabilities with intelligent agents and Web-based text and relating the work to proposed international data and metadata standards, such as American National Standards Institute X3L8 and International Organization for Standardization SC14.

SUMMARY

The potential for distributed hypertext as a means of managing and sharing environmental information is enormous. The Web provides clues as to what distributed information systems of the future can offer, but by no means can it be said that the Web itself is a solution to cooperative information management. Great strides remain to be taken in support of collaboration and intelligent information retrieval.

The EDEN Project is an ambitious effort to demonstrate the use of an agent-based system. As an application of the InfoSleuth intelligent agent technology developed at MCC, it represents a significant step forward in the potential to organize, access, and analyze environmental information. The collaboration between the participants is stimulating the development and adoption of appropriate data standards and methods for describing data elements that reaches well beyond the EDEN Project. If the pilot demonstration is deemed a success, it is anticipated that the InfoSleuth technology utilized in the demonstration will become commercially available.

ACKNOWLEDGMENTS

InfoSleuth2 was an MCC consortial research project sponsored by General Dynamics Information Systems (formerly Computing Devices International), NCR Corporation, Schlumberger, Raytheon Systems, Texas Instruments, TRW, and the DOD Clinical Business Area. This work was partially supported by NIST contract 50SBNB6C9076. The authors thank Mike Minock, Malcolm Taylor, and Vipul Kashyap at MCC for their contributions to this work.

NOTE

[1]InfoSleuth was an MCC consortial research project sponsored by General Dynamics Information Systems (formerly Computing Devices International), NCR Corporation, Schlumberger, Raytheon Systems, Texas Instruments, TRW, and the DOD Clinical Business Area.

REFERENCES

Bayardo, R.J., W. Bohrer, R. Brice, A. Cichocki, J. Fowler, A. Helal, V. Kashyap, T. Ksiezyk, G. Martin, M. Nodine, M. Rashid, M. Rusinkiewicz, R. Shea, C. Unnikrishnan, A. Unruh, and D. Woelk. 1997. InfoSleuth: agent-based semantic integration of information in open and dynamic environments. Pp. 195–206 in Proceedings of SIGMOD 97, Phoenix, Arizona, May 1997. New York: ACM Press.

Huhns, M., N. Jacobs, T. Ksiezyk, W.M. Shen, M. Singh, and P. Cannata. 1992. Enterprise information modeling and model integration in Carnot. In Enterprise Integration Modeling: Proceedings of the First International Conference. Boston, Mass.: MIT Press.

Nodine, M., and W. Bohrer. 1997. Scalable Semantic Brokering of Agents in InfoSleuth. Paper presented at the 18th International Conference on Distributed Computing Systems, Amsterdam, The Netherlands, May 26–29, 1998. New York: ACM Press.

Nodine M., and W. Unruh. 1997. Facilitating Open Communication in Agent Systems: The InfoSleuth Infrastructure. In Proceedings of the Fourth International Workshop on Agent Theories, Architectures, and Languages, Providence, R.I., July 1997. New York: Springer-Verlag.

Public Access to Environmental Information

PATRICK D. EAGAN, LYNDA M. WIESE, and DAVID S. LIEBL

Access to relevant environmental information can help to improve the ecology of industry. Imagine geographic databases that allow cross correlation of municipal and industrial discharges in various media (air, water, land). Envision the siting of industrial facilities on the basis of environmental carrying capacity of a geographic location. Think of the capability to determine the environmental profile of common industrial products. Technically, these capabilities exist. The ecology of industry can be improved by aggregating, evaluating, and increasing access to environmental information using information technology.

Indeed, there are several reasons to step up the aggregation and dissemination of information:

- There is increasing pressure to deliver public services more economically and effectively.
- Public agencies benefit by making issues public and engaging the public in decision making.
- Public-interest groups are demanding aggregated environmental information to monitor company performance and influence public policy.
- The demand for facility-based environmental information is increasing.
- Access to environmental information, such as the Toxics Release Inventory (TRI), has been shown to be effective in raising awareness about and reducing toxic chemicals.

The authors dedicate this chapter to Lynda's husband, Jeff Edge, and their daughters, Julianna and Lindsey.

- There is increasing demand in the private sector for facility-based environmental information for supply-chain management.
- Cross-media, aggregated environmental information will support better environmental decision making.
- The technology exists to deliver aggregated, spatially connected environmental information.

Public environmental protection agency administrators and industrial environmental managers, however, face fundamental questions about the availability of and, indeed, whether the collection of various types of environmental information will contribute to better environmental decisions in the agency or business. The public administrator's concern about accessibility and perceived value may revolve around whether greater public access to aggregated environmental information will lead to improved public policy, whereas the industrial manager may question whether access to aggregated environmental information will increase the ability to control risk from a supply chain, gather life-cycle environmental information for design purposes, or allow unwanted access to competitive engineering processes.

This paper discusses the value of aggregated environmental information to Environmental Protection Agency administrators and industrial environmental managers. It also presents the challenges in providing public access to this type of information, using the case study of an environmental information system called Fact, which is being implemented by the Wisconsin Department of Natural Resources (WDNR).[1]

AVAILABILITY OF ENVIRONMENTAL INFORMATION

Environmental regulatory agencies are mandated to protect the environment. As part of this mandate, federal and state environmental protection agencies collect large amounts of permit discharge and ambient monitoring data to assess regional and local environmental conditions, to determine compliance, and to charge fees. Industrial entities similarly collect vast amounts of environmental information related to their operations. Companies that implement environmental management systems under the International Organization for Standardization 14001, and adopt the goals of organizations such as the Coalition of Environmentally Responsible Economies or the Global Environmental Management Initiative, collect and publicly disclose much more information than is required by regulation. Each entity—industrial or governmental—collects environmental information to meet specific needs. The information, therefore, is available, but this information alone may not be what is needed to answer a specific environmental question or issue.

VALUE OF COLLECTING ENVIRONMENTAL INFORMATION

State agencies have partnered with other stakeholders to collect, aggregate, and analyze data. The WDNR (1996) report, *Acid Deposition Monitoring and Evaluation Program*, demonstrates the value of collaborating and coordinating efforts to meet data needs. In the case of acid deposition, the report states:

> Wisconsin's Acid Deposition Monitoring and Evaluation Program, as developed by the Acid Deposition Research Council, began its efforts during 1985 when the Acid Rain Law, Wisconsin Act 296, was enacted. In addition to its significant contribution in the area of acid rain research, the Acid Deposition Research Council model brought together diverse interest groups and then maximized the use of pooled resources to accomplish a common goal—to assess the threat of acid rain to Wisconsin's resources. Using the consensus approach, the Council has been able to reach agreement on complex issues like research objectives, priorities, and funding levels. Under Council leadership, the once acrimonious acid rain debate was transformed into an objective evaluation of facts.

Industrial environmental data also has proved to be especially useful for strategic planning by regulatory agencies and providers of state manufacturing assistance. For example, state agency personnel have correlated industrial emissions information from the TRI[2] with manufacturing processes. The TRI requires that companies report releases of toxins to the air, land, and water. Using the emissions data correlation, decision makers then set statewide priorities for reducing emissions at the source (Liebl, 1991, 1992). Community groups have similarly used aggregated environmental information to create demographic maps with toxic emission overlays as a basis for promoting environmental justice and to stimulate citizen activism to pressure industry to reduce the use of toxins (Dorr et al., 1993).

There are industrial and regulatory agency benefits to aggregated reporting of environmental information. The development of the Integrated Toxics Reporting System (ITRS) in Wisconsin allows for the identification and correction of data errors relating to hazardous waste generated by specific companies. The ITRS reports are reviewed by companies and cross checked against internal records for verification of data accuracy. The data from the ITRS also have been used by companies to develop full facility profiles as a first step to understanding the scope of hazardous waste generation and toxic emissions. These profiles offer managers incentives and justification to reduce toxins (WDNR, 1995).

GOVERNMENT DEMAND FOR ENVIRONMENTAL INFORMATION

In addition to the needs of public agencies for specific environmental data and information, there are several forces driving state agencies and the private sector to increase public access to environmental information. These include the need for greater administrative efficiency, the increased public demand for information, and the potential for improving environmental protection.

Administrative Efficiency

Public agencies have been under mounting pressure to decrease costs and increase productivity in monitoring industrial environmental performance; state agencies collect information (some of which is publicly available) from companies. This information is used to fulfill requests for paper copies of reports and permits or to provide notices of violations relating to specific companies. An environmental information system being implemented by the WDNR is expected to decrease the costs of copying, searching, and integrating relevant information; to speed delivery of information; and to ease records retention problems at the agency. A new data reporting protocol being tested would allow companies using the American National Standards Institute (ANSI) X-12 data standards to access the WDNR computing systems directly and submit the required data online. In fact, the future may bring direct links with company data systems such that the state agencies would not require annual reporting. Rather, state regulators might be provided with direct access to specific company environmental performance data sets that the company would routinely update for regulators.

The costs of maintaining and retrieving publicly requested information, however, can be substantial, even if done electronically. When facilities submit paper reports to state agencies, data from the report are entered into the database. Then it is stored and manipulated for billing and reports. Data entry operations cost time and money. Quality assurance for accurate data adds to the costs. For example, data comparisons with previous years' submittals are made to flag wide variations in data points so that agency engineers can investigate reasons for the wide discrepancies.

Electronic submittal of facility reports entails entering the data onto a computer diskette that is sent back to the state agency. As with paper-based reporting systems, resources are needed to support an electronic reporting system. For example, resources are needed to help first-time users with installing necessary programs as well as uploading and downloading information specific to each facility onto diskettes. In addition, technical staff is still needed to undertake quality assurance checks. However, responsibility for data entry errors rests with the company, not the state agency.

INCREASED PUBLIC DEMAND FOR INFORMATION

The mere existence of technology to increase public accessibility to environmental information has increased demand for information collected by state bureaucracies. Customers of state agencies (the public) are demanding that compliance and permit data be made available in electronic formats and be accessible on the World Wide Web (WWW). The spread of digital technology also has increased the demand for real-time information. Geographic information systems (GISs), which enable the collation, manipulation, and integration of spatial environmental information in ways not envisioned before, are being demanded by industry, government, and the public. This type of information can be used to show people the impact of releases from a facility or group of facilities. The GIS enables a visual map of an affected area to be overlain with a street map of the immediate area to give the public information about public risk.

Improved Environmental Protection

The reduction in emissions from industrial sources attributed to the TRI program has demonstrated the value of having a variety of environmental data compiled in one report and made widely available. The TRI requires certain facilities over a specified threshold to report releases of contaminants to the air, land, and water. This information is gathered annually and published for the public to review. Carol M. Browner, former administrator of the U.S. Environmental Protection Agency (EPA), observed that arming the public with basic information about toxic chemicals in their communities is among the most effective, common sense steps to protect the health of families and children from the threats posed by pollution. She pointed out that since the inception of the Community Right-to-Know program (under which TRI was implemented), reported releases of pollution into the community have declined by 46 percent (Kearns, 1997).

State agencies also have found TRI data particularly valuable in validating reports of various pollutants. For example, Wisconsin's air management staff have used their ITRS to identify the failure to report air emissions to the state's annual air emissions inventory, even though those emissions were reported under TRI. In the same way, Wisconsin also has begun using the ITRS at industrial facilities to ensure the quality of their own multiple reporting requirements and to ensure that all reporting is done accurately. Here the challenge is to use the data being collected to integrate single-media (air, land, or water) pollution reports, so that cross-media analysis can be done to help state agencies target companies that need assistance to minimize transfers of pollutants from one media to another.

INDUSTRY DEMAND FOR ENVIRONMENTAL INFORMATION

Aggregating environmental data enables more sophisticated analysis of environmental impacts of products and services and better environmental management and reduced risk. Progressive industry is interested in both product assessment and supply-chain environmental compliance performance.

Product or service assessments are based on the systematic, environmental review of the life-cycle steps associated with the product's component materials and manufacturing processes. Comprehensive analyses are called life-cycle assessments (LCAs). LCA has been used generally on products comprising few materials (Ehrenfeld, 1997; Owens, 1997). Aggregation of data across the product life cycle of energy and materials is a major feature of an LCA. However, there is a limit to the level of aggregation that is meaningful. Accumulating emissions from many different sites can lead to inappropriate conclusions if not integrated with other assessment techniques (Owens, 1997). Less-data-intensive, streamlined, or abridged analyses also are described in the literature. Several industries have effectively used streamlined matrix analyses on complex products (Graedel et al., 1995), processes (Eagan and Weinberg, 1999), and materials selection (Allenby, 1994).

Companies are also increasingly interested in the environmental performance of companies in their supply chains. Because of the consolidation of suppliers and a growing interdependence on suppliers and manufacturers to produce products and deliver services, the shared risk from environmental noncompliance or safety problems within supplier chains has become an issue for companies. Companies are therefore auditing suppliers for environmental risk (Anderson and Choong, 1997). Environmental performance data gathered on a supplier's compliance history and current status by state agency can be used by companies to estimate the quality of suppliers and judge the risks associated with those facilities.

CHALLENGES AND ISSUES RELATED TO IMPLEMENTING ENVIRONMENTAL INFORMATION SYSTEMS

Implementation of environmental information systems has not been easy despite the availability of technology. The implementation of WDNR's Fact System (see Box 1) revealed issues related to administration, public accessibility, and the interpretation of information.

Administrative Issues

Many administrative issues are associated with implementing a state-run environmental information system. The most obvious relate to designating responsibility for managing and administering the system. In the case of WDNR's Fact System, the small size of the databases envisioned suggests that it be run by the state even if some of the data are required for federal reporting. A problem that may arise from this approach occurs when the federal government or an industry wishes to integrate data from various states, but each state has a slightly different program with its own integration problems. With each state implementing its own system, data integration problems at the federal level are very likely to occur.

An anecdotal example in which one of the authors was involved demonstrates the next level of responsibility that relates to the data. In 1997, an environmental

---

BOX 1
The WDNR Fact System

The goal of the Fact System is to develop and implement an agency-level information system to support whole-facility pollution abatement work. The electronic database system is called the Fact System to signify that it will provide access to integrated facts about industrial facilities with environmental impacts.

The project has three phases: During the first phase WDNR built a facility environmental site register (ESR), the program's core integrating system, and a data warehouse from the ESR and the ITRS. In the second phase WDNR implemented WWW access to the Fact System. The third phase will add other information to the data warehouse, which may include compliance information, permit conditions, and linkages to the GIS.

Wisconsin is in the process of establishing 35 service centers across the state to bring most, if not all, department services closer to the public. Currently, 25 are operational. Each center is within 30 miles of a state citizen. The centers will offer public access to the Internet. These centers also will have computer access to the Fact System to help WDNR staff in decentralized locations make decisions locally that are part of a larger integrated ecosystem-based management of natural resources. Integrative tools, such as the Fact System, are vital to the staff in the service centers, who will be responsible for managing resources, for example, across a particular watershed.

The Fact System can also be vital to information sharing among the service centers. Currently, specific information on a facility could be in paper files in multiple locations—a WDNR central office in Madison, five regional headquarters, or the 35 service centers. The Fact System would allow access to basic information that is stored at a particular site from any of the WDNR offices, reducing duplication of information and the need to maintain multiple files in multiple locations. Because of their proximity to the people of the state, service centers also will allow every citizen access to WDNR's data systems along with technical on-site assistance to help them gather and interpret the data that they need.

URL http://www.dnr.state.wi.us/org/caer/cea/projects/one_stop/updates/overview.htm

---

group in Wisconsin accessed ENVIROFACTS, an EPA public information tool on the Internet, which contains information from environmental reports made by facilities to the EPA. The environmental group proceeded to make a Freedom of Information Act request for Discharge Monitoring Reporting System information for all pulp and paper facilities in central Wisconsin. The EPA complied with the request, indicating that reports of the past two quarters from one mill were not in the system. The environmental group proceeded with a press release criticizing the pulp and paper mill and hinting at violations of the Clean Water Act for the missing submittal. In fact, the reports had been submitted to the state authority, but the EPA files had not been updated.

This example raises the question of where responsibility lies to ensure that publicly accessible data are up to date, of good quality, and used responsibly. In addition to being concerned that the data accessed may not be current, businesses also are concerned about putting incomplete data in an easily disseminated digital form when they submit the data to the regulatory agency. The public expects publicly available information, whether digital or paper, to be accurate. One might expect that it is the responsibility of the data providers to ensure that the information is accurate, that data keepers will be responsible for how the information is compiled, and that the user is responsible for reasonable interpretation.

There is growing concern about the possibility that individuals who have outside access to important databases will tamper with the data. Two options often mentioned are: (1) using a redundant database exclusively for public access or (2) using a firewall[3] to protect the data and allowing read-only privileges. The use of redundant databases may be subject to uncertain financial support and the data currency issue described above. Hence, the use of firewalls is becoming common practice.

Finally, there can be data compatibility problems. Aggregating information from a number of different sources requires compatible data formats. For example, data on a single company that is derived from multiple regulatory reporting systems may be expressed in unrelated units of measure; denote toxins as elements, compounds, or mixtures; or list the facility by physical address or mailing address of a parent company. These inconsistencies must be reconciled prior to aggregating the data if reliable information is to be derived.

Public Accessibility Issues

Technical information traditionally has been delivered in writing or through personal contact. More recently, other alternatives have emerged to deliver technical information to as many people as possible using minimal resources. These methods include telephone conferencing, television broadcasts, videotape distribution, and automated faxing systems. The growth of the Internet has added electronic delivery of text through e-mail, file transfer protocols, gophers, and Web sites to the menu of options for transferring information. In addition, the use of e-mail listservers (where notices are sent to subscribers) and online database search engines has dramatically increased the availability of technical information to a wider audience. In the environmental protection field, this audience includes waste generators, regulatory agencies, technical assistance providers, and the public.

Electronic sources of environmental information, like other information, come in a variety of configurations. There are manually distributed databases for use on personal computers, interactive online databases that lead a user through a series of decision trees to locate information, online engineering and environmental library catalogs, and electronic information exchanges, such as e-mail listservers.

As states and the federal government move toward public access to electronic environmental data, questions arise about the technical capability of all citizens to access and interpret this information. While most public libraries provide access to the Web; finding and interpreting data can be a substantial barrier to the public. Public and private organizations are addressing this problem by indexing sources of information, compiling summary information on complex issues, and providing interpretation of environmental information (though often with a slant toward their own agendas).

The case for public access to environmental data is hard to dispute. No regulatory agency wants to say that their data cannot be made public; they are public agents empowered to enforce the laws. The debate over public accessibility revolves around how accessible the information is to the average citizen.

Wisconsin has an open-records law. Basically, all records reported to the WDNR, including environmental information, are open unless they meet the exclusion statute as a trade secret. By statute, there are two tests for a trade secret: (1) Has the company taken reasonable precautions to keep this information confidential? and (2) Would the release of the information cause the company to be put at a competitive disadvantage? Companies can request that certain information reported to the WDNR be kept confidential. Process-specific data, chemical usage, and production levels are the most common types of information protected as confidential. Much of the information gathered in annual reporting for the Air Emission Inventory includes company reports of production levels from which emissions are calculated using a standard set of emission factors—EPA air pollutant emission factors (AP-42). This causes concerns for highly competitive industries. Now that this information is easily accessible in electronic format, competitors can more easily gain intelligence that may influence pricing and competitive bidding.

Although environmental information always has been available in paper files or on microfiche, the authors are not aware of any evidence that an industry has taken the time to search through regulator's files to gain information on competitors. However, as profit margins shrink and competitors become more electronically adept, industries may look for any easily accessible competitive edge, such as evaluations of waste streams to reveal production processes or efficiencies.

Whenever the issue of public access to environmental data arises, an electronic data system is the first solution considered and the first funded. The public supports electronic data systems that are being designed by the government for the public's use. However, the average person needs to have the tools and training to be able to take advantage of these electronic data systems. Little research to date has explored how to raise the capacity of the public to embrace these electronic systems.

A fundamental problem with public access remains: Does the government collect environmental data that the public wants or can understand and meaningfully interpret? Local citizens want to know if they and their children can safely live near a manufacturing facility. Simply knowing that a manufacturing facility emitted 10 tons of permitted discharges into the air does not meet their need. Studies have shown that the American public is not well versed in environmental issues and may have limited knowledge to interpret the information (The Roper Organization, 1990). Although the government collects information for regulatory purposes, it may not know what information the public needs to make an informed contribution to the policy dialog. In all likelihood, the government has never asked.

Another accessibility issue pertains to environmental justice. Citizens living in heavily industrialized areas may not have the educational background and technical skills needed to interpret environmental data. Governments need to understand the range of skills of their target audience and then tailor information systems (electronic or otherwise) to them. Education to develop the interpretive capabilities of the public is also a possibility.

Interpretive Issues

Any attempt to aggregate and interpret environmental information must cope with the variety of waste types generated by a large industrial community and with the different reporting requirements and metrics of state and federal environmental agencies and programs. For example, in southeastern Wisconsin, nearly 5,000 companies generate 100 million pounds of hazardous waste and toxic emissions each year, comprising more than 300 chemical and waste types.

These waste and emissions data are compiled by the following reporting systems: WDNR Resource Conservation and Recovery Act Annual Reports, EPA TRI Form R Reports, WDNR Air Emission Inventory, and WDNR NR101 Annual Water Discharge Summaries. The data are reported in the following units: total mass in pounds, mass in pounds of toxic constituent, or concentration in mass per volume of water or air. In most cases, the lack of comparability between the reporting metrics results in under- or overreporting of emissions when the data are aggregated for a single facility. Extensive analytical interpretation is required if meaningful information is to be derived (Liebl, 1992).

In addition to providing an accurate summary of environmental information, regulatory agencies and others can provide further data interpretation for specific audiences. Individual companies may want to know how their level of emissions compare to an industrywide benchmark. Citizen groups may want to see maps that show air emission distributions over populated areas. Policy makers may want to track progress in environmental performance over time. Each of these special needs requires substantial data manipulation and interpretive configuration, and clearly will cost money.

CONCLUSIONS

As technology continues to advance, so also will the information handling and analytical capabilities of companies, citizens, and public agencies to deal with the important issues related to the environment. The importance of making linkages between the environmental conditions and human activities continues to grow. The benefits of aggregating environmental information to support private and public decision making are certain. Making usable, environmental information accessible to the public while promoting the public's capabilities to interpret the information is a complex problem. Because access to aggregated environmental information appears to be valuable, the administrative and implementation barriers that prevent society from taking advantage of this technology must be addressed.

ACKNOWLEDGMENTS

The authors acknowledge the contributions of Tom Aten, WDNR; Nicholas Bouwes, EPA; Ken Brown, Minnesota Office of Environmental Assistance; and John Stolzenberg, Wisconsin Legislative Council.

NOTES

[1] Issues such as property rights, described by Branscomb (1985) and Cohen and Martin (this volume), or other legal issues surrounding the submittal of electronic data or the confidentiality of personal information are beyond the scope of this paper.

[2] The Environmental Protection Agency is authorized under the Superfund Amendments and Reauthorization Act Title 313 to collect and disseminate information on the release of toxic chemicals to the environment. These data are compiled annually into the TRI.

[3] A firewall is an approach to computer network security; it helps implement a larger security policy that defines the services and access to be permitted, and it is an implementation of that policy in terms of a network configuration, one or more host systems and routers, and other security measures such as advanced authentication in place of static passwords.

A firewall system can be a router, a personal computer, a host, or a collection of hosts, set up specifically to shield a site from protocols and services that can be abused from hosts outside the site.

The main purpose of a firewall system is to control access to or from a protected network. It implements a network access policy by forcing connections to pass through the firewall, where they can be examined and reevaluated.

REFERENCES

Allenby, B.R. 1994. Integrating environment and technology: design for environment. Pp. 137–148 in The Greening of Industrial Ecosystems, B. R. Allenby and D. J. Richards, eds. Washington, D.C.: National Academy Press.

Anderson, J., and H. Choong. 1997. The development of an industrial standard supply-base environmental practices questionnaire. Pp. 276–281 in Proceedings of the IEEE International Symposium on Electronics and the Environment Conference, San Francisco, Calif., May 5–7.

Branscomb, A.W. 1985. Property rights in information. Pp. 81–120 in Information Technologies and Social Transformation, B.R. Guile, ed. Washington, D.C.: National Academy Press.

Dorr, L., J. Jaimez, and J. Haberman. 1993. Get to Know Your Local Polluter—Profiles of Minnesota's Top 40 Toxic Polluters. Minneapolis, Minn.: Citizens for a Better Environment.

Egan, P., and L. Weinberg. 1999. An application of analytic hierarchy process techniques to streamlined life-cycle analysis of two anodizing processes. Environmental Science and Technology 33(9):1495–1500.

Ehrenfeld, J.R. 1997. The importance of LCAs—warts and all. Journal of Industrial Ecology 1:41–49.

Graedel, T.E., B.R. Allenby, and P.R. Comrie. 1995. Matrix approaches to abridged life cycle assessment. Environmental Science and Technology 29:134–139.

Kearns, D. 1997. Toxics Release Inventory Community Right-to-Know. Press release, May 20. Washington, D.C.: U.S. Environmental Protection Agency.

Liebl, D.S. 1991. Using the Toxic Release Inventory for process-specific targeting of technical assistance opportunities in Minnesota. Pollution Prevention Review (Summer):295–300.

Liebl, D.S. 1992. Industrial Pollution Prevention Opportunities in Southeastern Wisconsin. University of Wisconsin–Extension, Solid and Hazardous Waste Education Center, Madison.

Owens, J.W. 1997. Life-cycle assessment constraints on moving from inventory to impact assessment. Journal of Industrial Ecology 1:37–49.

The Roper Organization, Inc. 1990. The Environment: Public Attitudes and Individual Behavior. Storrs, Conn.: The Roper Organization.

WDNR (Wisconsin Department of Natural Resources). 1995. Southeast Wisconsin Toxics Reduction Project Report. PUBL-AM 153–95. Madison: WDNR.

WDNR. 1996. Wisconsin Acid Deposition Monitoring and Evaluation Program. PUBL-AM-216–96. Madison: WDNR.

Internet Global Environmental Information Sharing

JOSHUA KNAUER and MAURICE RICKARD

A new model of global information sharing has emerged with the rise of the Internet. This new model allows information to be shared by many and among many, leading to a many-to-many distribution pattern rather than the standard broadcast-type distribution in which information goes from one to many. In addition, the construction of the information arises from exchange of information rather than from an imposed hierarchy. This paper discusses the new model and demonstrates its use in the implementation of the EnviroLink Network.

THE OLD MODEL OF INFORMATION SHARING

The old model of information sharing is based on a one-to-many distribution. It is constructed and controlled from the top of a hierarchical organization (such as with ownership or in management). For example, in the model followed by traditional media, a writer or team of writers is under one or more editors representing the interests of owners, engaged in sending one or more messages to readers, listeners, or watchers. The information flows in one direction only—from the media outlet to its consumers. This is the case with print media, radio, and television in mass markets and in small markets.

The economics of mass media require that the information to be shared be carefully selected. The selection process usually is based on what will generate the greatest reader, listener, and watcher response—either the sensational and the scandalous (ideally delivered before other competing media outlets) or the scoop (the most highly valued information). In the "scoop" model, information is primarily a commodity to be traded, not to be shared. When it is delivered (to the media consumers, not competing media producers), its value as unique information is lost or traded with the hopes of a wider circulation, larger readership, listening audience, or viewership for more advertising revenue.

INTERACTIVITY UNDER THE ONE-TO-MANY MODEL

In the old model, information is shared with little expectation of feedback other than regular consumption—watching, listening, reading, or purchasing. The end result is more of a lecture than a conversation. Opportunities for interactivity do exist, but they are limited, inefficient, and far less responsive than those offered by the Internet.

Letters to the editor are a form of interactive communication under the old model of information distribution by print, but they are very inefficient as a medium of information exchange. This inefficiency takes several forms: time delays from composition to publication and from reading to response, space limitations, and limited number of response cycles in an interaction.

The delay from composition to publication ranges from a day or two (in the case of a local newspaper) to a month or more (in the case of a monthly national magazine). This includes time necessary to transport a physical letter (recently reduced by the increase of e-mailed letters to editors) and the lead time required for preproduction, printing, and delivery by the media carrier. Editorial decisions in selecting the letters for print act as an additional barrier to the exchange of information. These decisions can be based on the amount of available space or on the content of the individual letter. The limited space for letters results in relatively few letters being published (relative to the total circulation of the publication). Letters that are run often are edited to fit the available space or are selected for brevity. In this case, not every reader is an equal (or potentially equal) participant in the information exchange.

The number of response cycles in this model also is limited. Although one occasionally may find letters responding to other letters, these responses to responses rarely are printed beyond one iteration. A true conversation requires more response cycles in real time.

Radio and television also offer possibilities for interactivity. Like the print media, however, there are limits on the level of participation, based on available time rather than available space. In the case of radio and television call-in talk shows, the time slot of the show precipitates an inverse relationship between the number of participants and the time available for each one to participate, thereby eliminating dialog that requires longer discourse. Any media outlet offering real-time call-in viewer and listener interactivity inevitably will reach a state offering relatively few opportunities for participation. The format and program will become either too popular to give viable access to all of its listeners or viewers, or it will remain accessible only to a limited population. Even if a given show's distribution grows (from local to other markets to nationwide), each audience member's opportunity for participation approaches zero, because the time available for participation remains constant.

Even if a show does not become popular and grow into new markets, it may offer a greater possibility for participation to its small number of consumers, but it will serve as an information-sharing venue for a limited population, and perhaps face the threat of cancellation. The call-in survey form of interactivity circumvents the problem of time limitations by tallying simple yes, no, and undecided responses. This allows a greater number of participants in each show, but the limited response options constrain the expression of subtle or nuanced positions.

THE NEW MODEL: MANY-TO-MANY INFORMATION SHARING

Many-to-many information sharing is a conversation, not a lecture. It is a free interaction among many parties. In the new many-to-many model, every reader is a potential writer, every listener a potential speaker, and the cost of entry to the conversation is relatively low compared with that for print, radio, or television. The only requirement is access to a computer, a modem, and an Internet connection.

The Internet as a Medium for the New Model

Although Internet connections are not universal, most of the developed world and an increasing portion of the developing world have some access to it. Universities, schools, businesses, governments, and private individuals with computer equipment can have direct access to information from almost anywhere on the planet. In fact, people without computer equipment can gain the same access through libraries, schools, and the growing number of Internet cafes. For example, it costs as little as eight dollars for someone in midtown Manhattan to hold an hour-long conversation with someone in Nepal (Greenwald, 1997).

Unlike television, participants can talk directly to the original producer and other participants (not just the company that produces the shows). Unlike talk radio, the conversation is not limited to one geographic area. Unlike both television and radio, interaction on the Internet is not subject to the constraints of time. Unlike newspapers or magazines, the turnaround time for information sharing is nearly instantaneous, and interaction is not marginalized or limited by constraints of space. A constant global conversation is therefore possible.

Furthermore, the global conversation taking place over the Internet consists of millions of participants simultaneously sharing information on a great variety of different topics. The forum, in addition, is under no one controlling agent. There are no fixed rules. It is essentially a functioning anarchy. The Internet is currently the only example of a large-scale unregulated (or sparsely regulated) system in daily use.

Much like other media, anyone (with the necessary equipment) can be a consumer. The Internet, however, enables almost anyone to be a publisher as well, adding his or her voice to the global conversation. The increasing commercial use of the Internet notwithstanding, there are many niches (provided by educational and public institutions as well as nonprofit organizations) that facilitate individual or group expression free of commercial considerations or censorship. The great number of participants, the variety of utterances on the Internet, and the lack of overriding control of the system contributes to both positive and negative effects.

Positive Effects of the Internet

The Internet enables larger numbers of people to be brought into a dialog and to share and combine this information set into a larger, multifaceted knowledge base. Global challenges and problems (socioeconomic, political, environmental) are complex, multifaceted, and multidisciplinary. Solutions to such problems require extensive knowledge bases, made up of contributions from large numbers of participants with their own different knowledge sets and perspectives. For any given group, the greater the difference in specialties and knowledge sets, the less likely it is that the group might have engaged in any kind of dialog through channels other than those offered by the Internet.

The Internet also allows wider participation in the political process. The cost of entry to the global dialog on the Internet is much lower than, for example, the cost of entry to mainstream political participation. This allows the disempowered and marginalized a place in the global dialog, using the same distribution channels as used by dominant cultural forces.

In particular, for independent publishers the Internet has facilitated the sharing of information and the circumvention of possible censorship and repression. Even when governments or other interests pressure one or more Internet publishers to curtail or inhibit discussion of certain subjects, other publishers are likely to take on particular causes and create venues for continued discussions.

A noteworthy example of this relates to a report criticizing police investigations and government mishandling of allegations of child abuse in Nottingham, England. The report proved to be such an embarrassment that the printed report was suppressed by the British government. However, it was published on a Web site in England, which then was pressured to remove the report from its server. Within days, the report had been copied to several servers around the world, effectively circumventing attempts to censor a critical point of view (Craddock, 1997).

An Internet-based Albanian newspaper shows the use of the Internet to share information amidst censorship. Although access to the Internet is limited in Albania, reporters there e-mail their stories to volunteers in France, who then post the stories on the World Wide Web. This transnational team uses the

Internet to supply the world at large with the only uncensored news out of Albania (Nouzareth, 1997).

Finally, the electronic distribution of a vast amount of textual and visual information on the Internet can occur without any increase in consumption of natural resources (trees, chemicals used to process wood into paper, inks, and fuel expended to deliver printed material).

Negative Effects of the Internet

The negative effect of the Internet that is discussed most frequently is the increased access by consumers—children, particularly—to socially unacceptable, damaging, or objectionable information. Racist tracts, how-to guides for terrorist acts, and other troubling material are more universally accessible on the Internet than in printed form. Although objectionable material is certainly available on the Internet, it is also less prevalent than is suggested in reports by traditional media (Katz, 1997a,b; Rosenberg, 1997).

An equally obvious but persistent negative effect of the Internet as a new medium for information exchange has to do with the quality and quantity of information available. It is easy for information seekers to be overwhelmed by the information. One concern with users being overwhelmed is that they may become more passive in their interactions on the Internet. One media observer (Thomas, 1995) has identified this as the "info-sedative nature" of the Internet. Passivity is a particularly detrimental (and all too frequent) response to information about critical global environmental problems such as degradation of rain forests, famines, global warming, disease, ozone depletion, and other issues that benefit from the active participation of media consumers in generating and acting on solutions.

Dualistic Effect of the Internet

Concern about the quality of information available through the Internet revolves around the veracity of the information that customers receive. Critical consideration of the sources of information and their motives will help to separate valuable content from noise. This is particularly important because the users choose from hundreds of thousands of sources online. They also rely on information providers with whom they develop a sense of trust. The sources of information are no longer the familiar few that are part of the traditional media. For example, in the print media, consumers more readily can rely on the media brands (e.g., *The New York Times, The Washington Post, Nightline*) to determine how the information is filtered (conservative or liberal, environmental proponent or environmental conservative) and whether the sources can be trusted on the basis of their history of checking their sources and their facts. A story running in the *New York Times* or *The Washington Post,* for example, would be more credible with many readers than a story running in the *National Enquirer* (except for readers for whom the reverse would be the case).

On the Internet, it is more difficult to discern the reliability of information. It is possible for the information to appear official or veracious when it is not. For example, Pierre Salinger's dubious contributions to the debate over the cause of the TWA flight 800 tragedy was based on a fictitious "report" of a friendly-fire incident, which sounded like an official military report. What Salinger failed to do was to verify the source before passing the report on to others, which many reporters had done several months earlier, rejecting the report as a hoax (Rosenberg, 1996).

As with any information, but most particularly from Internet sources, critical analysis by consumers is essential. Assuming that the consumers are likely to do so and are capable of doing so, there is a likelihood of consumers engaging fully in the information-sharing process. There is evidence that when Internet users engage fully in verifying reports, even those reported by the more reliable press sources, a fuller picture of a particular situation can merge.

A case in point is found in the financial reporting of the Iomega Corporation (Gardner and Gardner, 1996). In 1995, the Iomega Corporation came out with a new disk drive whose disks held 70 times the capacity of existing floppy disks. Between the announcement and the debut of the drive (a matter of months), Iomega's stock price went from $2 a share to $10 a share. The rise drew strength from the expectations created by laudatory reviews from top computer industry publications that lavished praise on the new product. However, skeptics started to question the company's capability, pointing to the company's lack of manufacturing capacity and its two previous years of losses. Rumors abounded that the stock would shortly return to its initial offering price of $2. The reaction from private investors was not to dump the stock; quite to the contrary, in metropolitan centers across the United States, private investors began polling their local computer stores about their current stock and their backlogged orders of Iomega disk drives. They then went online with this information in a public discussion of Iomega availability. Simultaneously, an engineer took a simple tour of the company's facility and observed the manufacturing process. He then went online to contribute his estimates of Iomega's production capacity. Furthermore, another investor drove to the company's headquarters on Sunday afternoon in order to report on how many cars appeared in the company parking lot and then went online with his information. What resulted from the collection and online publication of these seemingly inconsequential details was a national public conversation, a conversation that had never been possible before. How well the company was able to meet demand—a subject of so much speculation among offline investors—had a sure answer among online investors. The information was provided so quickly and in such detail that no Wall Street analyst or firm could have done it. And in fact, within a week, Wall Street traders were becoming part of the discussion.

A single location on the Internet had become the place to go for understanding and valuing Iomega.

When information consumers think critically about the information they receive, they can engage the information on a much deeper level than is normally the case. The Internet provides a means for information consumers to participate fully in the exchange of information. By making their interpretations available to others, they become community resources of critical and analytical information, other voices brought fully into the global conversation. Each reader is no longer a potential writer but an actual writer. Each listener can speak. Information truly can be shared globally.

GLOBAL INFORMATION SHARING AS A FORCE FOR ENVIRONMENTAL CHANGE

All current environmental science points to the world as a vast system of relationships in which many small changes can accumulate to become significant global changes. Local environmental information needs to be shared worldwide in part because of its global implications, but also because observed environmental changes in one locality can be related to changes in another. If solutions to our growing global challenges are to be found, information will have to be shared quickly across distances, across disciplines, and across ideological divisions. At this time, the Internet offers the greatest opportunity for the growth of realistic, workable, global information-sharing systems.

The many-to-many model of global information sharing can be particularly useful for businesses, educators, government agencies, and others seeking information about sustainable and ecologically progressive practices. Businesses in particular can benefit—and have benefited—from engaging in the global environmental information-sharing dialog on the Internet. Businesses can connect with their peers within the sustainable business community to share tips, advice, and other information about sustainable business practices and strategies. A startup company could use the Internet to research other companies' progressive labor policies, materials, and methods used in sustainable business practices and find contact information for socially responsible vendors. In addition, the company could make its products and services available to a global base of customers through the Internet at a fraction of the cost of conventional advertising and promotion.

Government agencies can use the Internet to research examples of other governments' efforts at sustainable policies and practices. Government agencies also can use the information-sharing capability of the Internet to initiate and maintain contact with constituents, generating and sustaining a two-way data flow to provide daily updates on their activities and receive constituent feedback.

Community groups working on, for example, restoration and conservation projects can connect with other communities that have completed similar projects to learn how developers, community groups, and government have worked together to find solutions to environmental problems. At the same time, the results of their work can be made available on the Internet, adding to the global library of environmental information to be used by others.

THE ENVIROLINK NETWORK AS A MODEL OF GLOBAL ENVIRONMENTAL INFORMATION SHARING

The EnviroLink Network was built to facilitate the many-to-many environmental information sharing model on a global scale. Founded in 1991 as a nonprofit organization, the EnviroLink Network serves the global community as a central location for environmental information on the Internet. As of April 2001, the EnviroLink community consisted of more than 500,000 visitors (estimated) per month[1] in over 130 countries, on six continents, connecting to each other through EnviroLink's Internet-based services: a Web site, real-time chats, bulletin boards, and electronic mailing lists.

The EnviroLink Network provides, through its content areas and other services, a forum in which the global environmental conversation can take place. Built from the ground up, members of the global EnviroLink community are using environmental information not only for reference and research purposes, but also in political, artistic, and community applications. To reflect and facilitate this variety of uses, EnviroLink's resources for global information sharing are organized in several different content areas, which are described below.

News Service

The EnviroLink News Service runs daily news stories on environmental issues and events. It is read by students, educators, governments, organizations, ordinary people—anyone interested in environmental information that is updated daily. Its original reports, written by over 800 volunteers worldwide, are frequently reprinted in many different publications around the world.

The Sustainable Business Network

The Sustainable Business Network (SBN) is an area of the EnviroLink Web site that is targeted at a specific audience: businesses that follow socially responsible business practices and people interested in purchasing services and products that are made by these companies. The companies highlighted in the SBN produce items made from recycled materials, and promote organic farming practices, alternative energy sources, energy conservation, and other green products and technologies.

The SBN itself is built from global information sources. Its contributors are writers from around the world whose reports and articles are updated monthly. It features interactive bulletin boards for readers, writers, businesses, and customers to engage in follow-up dialogs about stories, job opportunities, exchange of ideas, partnerships, and other information.

Library

The EnviroLink Library contains links to thousands of environmental resources on the Internet organized in over 250 categories. Some of the resources are hosted on EnviroLink and some are hosted elsewhere. As the central resource linking environmental information on the Internet, the EnviroLink Library is an prime example of a global information-sharing tool.

EnviroArts

The EnviroArts area of the EnviroLink Web site represents a different form of global information sharing, based on the notion that the art of a society transmits unquantifiable information that can lead to greater understanding than that captured in scientific concepts. Through the Internet this information is made available to many more people than it otherwise could be. EnviroArts hosts the work of select environmental artists, and links to artworks with environmental themes and concerns elsewhere in the world. The EnviroArts gallery is located everywhere, available to anyone, and open 24 hours a day.

Examples of Global Information Sharing

Individuals, businesses, and groups everywhere in the world have used the EnviroLink Network to share information. For example, a story on Royal Dutch Shell's operations in Nigeria originally ran on the EnviroLink News Service content area of the EnviroLink Web site before being picked up and reprinted in an environmental newsletter in Latvia. Similarly, *Green Connections* (a permaculture publication based in Australia) found *Green Marketing and Management: A Global Perspective*—a book written by SBN content partner John Wasik—by using EnviroLink's SBN. *Green Connections* used Wasik's recommendations to environmentally restructure the functioning of their office. They analyzed their purchasing patterns and changed them to make them more sustainable, buying office supplies made locally by sustainable businesses instead of buying conventionally produced supplies from other countries. Their contact with John Wasik on the SBN prompted them to question their practices and investigate associated implications of their practices and to change to more sustainable and progressive practices.

Teachers from around the world regularly use the EnviroLink Library as a research tool for their students to find information for reports and projects. Currently a school district in New Zealand is using the library as an information source for their assignments and research projects.

Staff members from the office of Vice President Al Gore have used EnviroLink resources to research public response to administration policies, announcements, and legislation. The Congressional Research Service and individual staff from the offices of various senators and representatives regularly use EnviroLink for similar applications.

The EnviroLink Network is a working example of the new many-to-many information-sharing model. Constructed from the ground up, rather than as a hierarchy imposed from the top down, EnviroLink provides the opportunity for a global dialog in which everyone is a potential participant, able to contribute information, to listen, and to support the venue for the conversation.

NOTE

[1]Our figure for estimated visitors to the site is taken from a daily record of unique machine addresses requesting files from our site, assuming that one person uses each machine.

REFERENCES

Craddock, A. 1997. UK activist: let 1,000 mirror sites bloom. Wired News. *http://www.wired.com/ news/news/politics/story/4461.html.* [October 19, 1998].

Gardner, D., and T. Gardner. 1996. The Motley Fool Investment Guide. New York: Simon & Schuster.

Greenwald, J. 1997. K@mandu. Salon Magazine. *http://www.salonmagazine.com/march97/ news/ news970311.html.* [October 19, 1998].

Katz, J. 1997a. Media hysteria. Part I: epidemic of panic. Hotwired. *http://www.hotwired.com/ netizen/97/18/katz0a.html.* [October 19, 1998].

Katz, J. 1997b. Media hysteria. Part III: finding the cure. Hotwired. *http://www.hotwired.com/ netizen/97/18/katz4a.html.* [October 19, 1998].

Nouzareth, M. 1997. Albanian Daily News. Scripting News. *http://www.scripting.com/ frontier/ stories/albania.html.* The publication itself is available at *http://www2.AlbanianNews.com/* AlbanianNews. [October 19, 1998].

Rosenberg, S. 1996. Blame it on the net. Salon Magazine. *http://www.salon1999.com/ media/ media961112.html.* [October 19, 1998].

Rosenberg, S. 1997. Indecent exposure. Salon Magazine. *http://www.salonmagazine.com/ 21st/ straight07.html#e.* [October 19, 1998].

Thomas, D. 1995. Pere Ubu FAQ. Ubu Communex 33095. *http://www.dnai.com/~obo/ ubu/ ubu_FAQ.html,* #8. [October 19, 1998].

Knowledge Networking for Global Sustainability
New Modes of Cyberpartnering

NAZLI CHOUCRI

By a conservative count, there were more than 50 million Internet users worldwide in 1995, a figure that was projected to exceed 200 million by the turn of the century. In retrospect, this projection substantially underestimated the number of Internet users today. For example, although most current usage is concentrated in the United States, which accounts for about 50 percent of worldwide web use, more than 80 percent of web users are expected to be outside the United States. Far more compelling is the expansion of the e-economy and its impacts on the traditional economy. In 2000, it was estimated that the e-economy in the United States generated $830 billion in revenues; in January 2001, the e-economy supported more than 3 million workers. The "new" economy appears to be taking a notable place alongside the "old" economy.

This broad pattern is distinct from, but related to, another major global trend—the remarkable increase in worldwide concerns about and attention to environmental problems, coupled with concerns about transitions toward sustainability (Alker and Haas, 1993). This second trend is the direct result of a growing awareness that industrialization in the West has led to increasing, pervasive pollution of the environment. Environmental degradation was, and continues to be, an inescapable feature of the old economy. Thus, we are faced with a major global dilemma.

Broadly construed, the global dilemma is this: *If* the combined impacts of human activities and conditions, shaped by prevailing social norms and values, continue to place serious strains on life-supporting properties, threaten natural and social systems, and generate propensities for social conflict and violence, *then* the historical patterns of unrestricted growth that have been so successful in

195

the past can no longer provide reasonable guidelines for moving toward viable futures for the population of this planet (Choucri and North, 1993). And the broad-based consensus about the implications of the dilemma is this: *If* societies understood the extent of the dilemma, *then* they would engage in efforts to identify strategies to ensure their security, survival, and sustainability.

It is fair to say that there has been (and continues to be) a proliferation of efforts to arrest erosions in life-supporting properties that are directly traceable to human actions. The quest for sustainability is a distinctive feature of contemporary thought in both scholarly and policy circles. Even the business community is beginning to talk about corporate social responsibility in search of sustainability. Yet, few coherent frameworks have been developed to guide decision and policy making. The development of effective strategies has been impeded by some major obstacles, both in theory and in practice. First, considerable ambiguities remain about the meaning of "sustainable development" and the conditions necessary for the viability of natural and social systems (Lang, 1994; Rothenberg, 1993). Second, there has been an explosion of all types of information of varying quality and reliability, as well as considerable difficulties in tracking the significance of information or follow-up actions (see, for example, Tolba et al., 1992). Third, technologically, even relatively simple concepts such as global conferencing have not been translated into routine practices. Finally, information and knowledge exchange has been limited by infrastructure disconnections and limited feedback.

None of these obstacles alone is insurmountable, but together they have severely impeded the effective use of advances in information technologies and their deployment for knowledge-sharing and management. They have also impeded the exercise of political will for acting on emergent knowledge about environmental degradation or addressing a wide range of environmental problems or moving towards transitions to sustainability. These obstacles are all independent in origin, dynamics, technological foundation, and policy priorities, but they appear to be converging in ways that may lead to a paradigm shift in our understanding of how knowledge-based uses of advanced information technologies can be effectively used on a global scale.

If there is a serious commitment to sustainability as an alternative to the conventional economic growth model, then the search for—or development of—a sustainability model must draw upon relevant knowledge and evolving interpretations of collective experiences, best practices, new theories, innovative technologies, and new social modalities—in all economic sectors, all geographic regions, and at all levels of development (Choucri, 1993). This paper introduces the Global System for Sustainable Development (GSSD), an Internet-based knowledge-networking system predicated on an internally consistent framework for organizing knowledge and for guiding action pertaining to the broad domain of sustainability. As a distributed system, GSSD combines the power and resources of the Internet with new strategies for knowledge-sharing on a global basis (Choucri, 1995).

GSSD also serves as the core platform for the Global Partnership on Cyberspace for Sustainability, which was introduced at the Fifth Session of the United Nations Commission on Sustainable Development, May 1997, and subsequently at the Special Session of the General Assembly, June 1997, known as "Earth Summit +5." At those meetings it was suggested that a viable partnering platform on a global scale would require the following characteristics:

- effective representation of the knowledge base related to sustainability
- efficient electronic capabilities that represent the frontier of advances in information technology
- mirror-site options in locations overseas and in languages other than English

By providing new venues for cyberpartnering, GSSD transcends conventional, disciplinary foci and encourages interdisciplinary analysis and understanding. It provides guidelines for transitions from centralized management to distributed knowledge generation and knowledge sharing. It also facilitates a change from conventional web posting to interactive, adaptive knowledge exchange and e-conferencing. The core concept of GSSD is sufficiently broad to include both the scientific tradition and the more policy-based, pragmatically oriented traditions of business and industry and governments.

THE GLOBAL SYSTEM FOR SUSTAINABLE DEVELOPMENT

The GSSD is an adaptive, interactive system for knowledge networking, knowledge management, and knowledge sharing for use in conjunction with Internet resources. Its goals are (1) to define the dimensions and dilemmas of changing from current policies and strategies based on imperatives of growth, (2) to identify policies and strategies that facilitate social and environmental sustainability, and (3) to track the range of policy responses nationally and internationally. The GSSD knowledge base is organized as a hierarchical, embedded system of topic-specific, cross-indexed, content-rich Internet resources in the following areas:

- human activities and conditions
- sustainability problems associated with human activities
- scientific and technological solutions
- economic, political, and regulatory solutions
- evolving international actions and responses

Since the GSSD knowledge base is evolving, dynamic, and adaptive, its content changes as new evidence and insights are developed or as new theoretical or policy perspectives emerge.

Basic Features

GSSD can be described as an integrated knowledge system consisting of a coherent, conceptual framework; a transparent, knowledge-sharing strategy; specific product-oriented applications; and distributed updates, streamlined maintenance, search and browser facilities, and multilingual functionalities and mirror sites. GSSD operates as a metasystem (i.e., a network of networks). The GSSD knowledge base is an evolving collection of topic-specific abstracts of Internet materials, prescreened for content and quality, cross-referenced, and indexed; a link is provided from each content-specific abstract to the web site for the original source. In other words, the value-added of the GSSD knowledge base is its content, quality, coherent framework, *and* its e-connections to the original Internet materials. The selection and inclusion of sites is subject to systematic procedures.

The integrated design for the GSSD knowledge system is derived from a dynamic, multifaceted, interdisciplinary, and international approach to the domain of sustainable development (Becker and Jahn, 1999; Choucri, 1999). GSSD is structured to accommodate updates of knowledge, changes in intellectual orientation, and concurrent, disparate perspectives, as well as diverse information systems and structures. Designed in embedded and hierarchical terms, GSSD consists of "rings" representing nested substantive crosscutting news into domains of knowledge (Figure 1). The structure of the system is informed theoretically by ongoing research about how patterns of behaviors in international relations can be traced to interactions among variables in populations, resources, and technologies (Choucri and North, 1975, 1989, 1993). The operation of the system as a whole does not depend on a particular perspective or on the underlying research or evolving findings. The GSSD core is a "placeholder" for current assumptions, with the understanding that they might be revised in light of new knowledge, better analysis, and overall improvements in the quality, scale, and scope of our understanding.

The design structure for the substantive content (Figure 2), which is formulated in terms of subject or topic "slices," takes into account sector-specific economic activities (Figure 2, top) as well as broad sociopolitical domains of human activity (Figure 2, bottom). These multiple perspectives are important because subject slices are usually examined on a stand-alone basis without reference to interconnections; and expertise, by definition, follows this conventional specialization (or fragmentation) of knowledge, often undermining the potential for integration and interconnections. Table 1 lists key features of the overall knowledge-base system spanning the 14 slices and the 5 rings.

Knowledge Strategy

As indicated above, the GSSD platform consists of a system of detailed, internally consistent characterizations of the contents of the key dimensions of

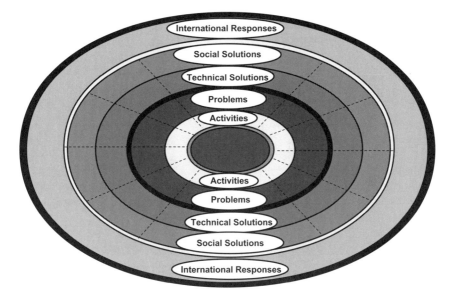

FIGURE 1 Dimensions of sustainability framework: rings or perspectives. SOURCE: MIT, 2001.

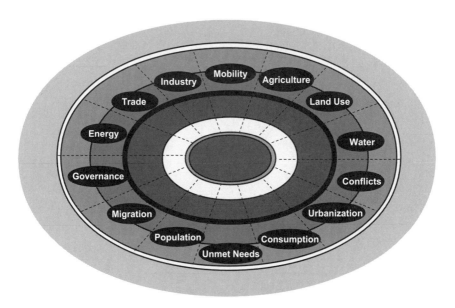

FIGURE 2 Structure of knowledge system: slices or topics. SOURCE: MIT, 2001.

TABLE 1 GSSD Knowledge Base: Illustrative Features

Type:	Metric and nonmetric
Subject:	Concepts, theories, cases, agreements, indicators, models, etc.
Status:	Private, public, mixed
Uses:	Preset, or customized, for search, browser, and navigation options
Content:	Cross-referenced, indexed abstracts of Internet resources
Structure:	Hierarchical, nested, system design
Coverage:	Selective and quality controlled; decentralized knowledge provision
Languages:	English base, with Chinese, Arabic, French and shortly Italian, Japanese, and Spanish operations. If original entries are not in English, materials are provided in all supported languages.

sustainable development, defined in terms of individual topics (slices) and their attendant content (rings). The system has four rings across all slices and one ring that transcends all of the slices or topics. The four rings common to all slices address (1) a topic or activity, (2) problems generated by a particular activity, (3) the technical and scientific solutions to such problems at any point in time, and (4) the social, economic, and regulatory solutions available to date. These details are shown in Figures 1 and 2. Each of the 14 substantive dimensions for the broad domain of sustainable development is further differentiated in terms of a detailed "slice outline," which refers to the full-blown characterization of the contents of each subject (or slice).

The knowledge content of individual topics, namely, the slice outline, is organized as a set of conceptual, as well as practical, functions. Each slice outline consists of concepts and subconcepts that are used as "tags" to index individual items in the knowledge base. These tags enable characterization of current knowledge so that updates can be made easily; provide guidelines for populating the system, namely, the selection of prescreened, "spidered" web sites; and ensure a certain degree of internal consistency in the development of the GSSD knowledge system in terms of the selection of indicative web sites. (In a different idiom, the slice outline is akin to a table of contents combined with an index for a printed book. It can also be thought of as a subject-based directory.)

If we consider the 14 topic-specific slices as reflecting the domain of sustainable development—problems and potential solutions—then the outer ring of GSSD (the fifth ring) is not connected to any specific slice. It transcends all of the substantive issues (i.e., all slices) thus representing access to knowledge (data, policies, actions) pertaining to different types of coordinated international responses and global accords, large-scale policy measures designed to address sustainability-related problems. Figure 3 shows the contents of this ring.

FIGURE 3   Global sustainability strategies: types of coordinated action.   SOURCE: MIT, 2001.

The value of the GSSD network of networks is enhanced by indexed connectivities across different sustainability-related issues and domains of human activity. Its "metafeature" is defined by the strategy of networking among networks. Hence, the knowledge strategy is fundamentally one of networking, sharing, generating synergism, and building in correctives—in the sense that the GSSD knowledge base evolves over time, drawing on a wide range of Internet resources and information systems.

Key Applications

An overview of GSSD applications and capabilities is accessible from the GSSD home page at the "Introduction" button, and a more detailed review is presented in the individual applications buttons. For example, Figure 4 shows a screenshot "Using GSSD." All applications are based on the assumption that users will interact with the system in one (or more) of the following modes: an *access user* (to obtain knowledge types, or basic data, through search options and navigation tools); a *knowledge provider* or input user (to place contents of pre-screened web sites in the system's knowledge base); a *knowledge developer* (to enable the organization of local knowledge and its formatting for use in global

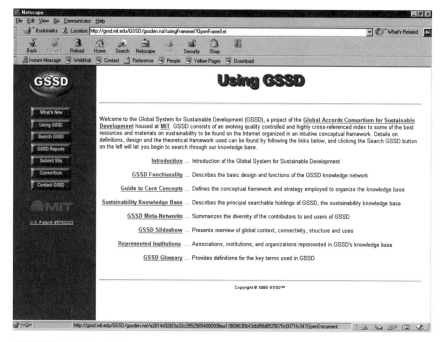

FIGURE 4   Using GSSD: highlights of system functions. SOURCE: MIT, 2001.

networks); or a *multimode e-user* for wide-area connectivity, knowledge management, and networking. The multimedia mode, currently in experimental form,
will enable a user to network through audio and video facilities and, more important, to engage in real-time (almost laboratory conditions) survey research on a
worldwide basis. For many social science and most policy-relevant uses, survey
applications of this type would provide unparalleled opportunities for accessing
and measuring select groups in targeted modes or the global community as
a whole.

System Access, Navigation, and Search

Access to the system is obtained via icons representing intersections of slices
and rings. Clicking on an icon takes a user directly to a requested list and brief
synopsis of Internet sites. After clicking on the cell of interest, the user has direct
access to that site and essentially exits GSSD. This application can be characterized as passive use of GSSD. (A text-based version of GSSD is currently under
development.)

Users have six options for exploring or drawing upon the GSSD knowledge base. Two of these are in conventional search mode, (1) a simple search

(text search) or (2) an advanced search (reflecting specific requirements). The four other options involve more detailed access or search strategies: (3) selection by slice, (4) selection by ring, (5) selection by concept, or (6) selection by cell (i.e., a more fine-grained or detailed feature of a broader concept). The search and navigation options operate over the entire GSSD and can be used with a high degree of specificity for targeted segments of the data. The screen shot in Figure 5 shows four of the six "search types" at an aggregate level.

Knowledge Management

The system input application decentralizes the tasks of knowledge management, maintenance, screening, and quality control. This function is currently performed by the GSSD system administrators but is intended for distributed use in collaborating institutions. Automated input and update capabilities will eliminate the need for any programming on the part of the user. As shown in Figure 6, only a few items of information are required as user inputs to automatically update the system. This feature will facilitate data entry and hence facilitate implementation of decentralized capabilities.

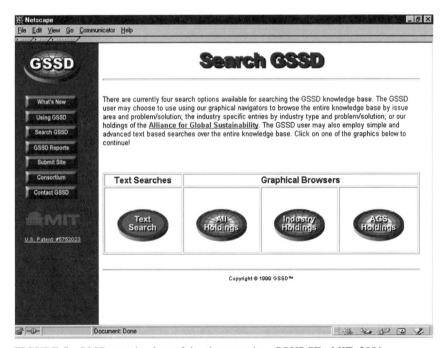

FIGURE 5 GSSD search: four of the six strategies. SOURCE: MIT, 2001.

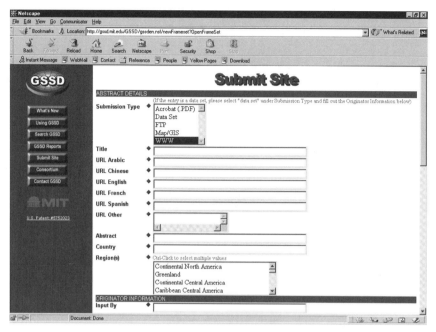

FIGURE 6   A GSSD submit site. SOURCE: MIT, 2001.

Multilingual Functions

A recently developed feature of GSSD—the provision of multilingual functionality—is especially useful in a world that is increasingly e-connected. The
Internet today is an English-speaking medium in a non-English-speaking world;
cyberspace is increasingly populated by non-English speakers. The multilingual
capability of GSSD involves a multiple-language interface and workflow that
currently includes Arabic, Chinese, English, and French. Current GSSD multilingual functionality enables:

1. **Improved access to knowledge.** Reducing the difficulties facing non-
   English speakers will enable them to find specific items or information on
   the Internet through retrieval of abstracts. The abstract (description) of
   each site in the GSSD knowledge base is translated into all of GSSD's
   supported languages. The translated abstracts are then available for
   e-searches through the system's six search modes.
2. **Strategic use of resources.** GSSD's abstracts will inform the user in
   advance of translation where the most fruitful information is housed, thus
   significantly improving access and efficiency.

3. **Expansion of knowledge base.** A platform for non-English content enables non-English-speakers to make their own data widely available and increases the amount of local knowledge on the global network.

Knowledge Networking and Strategic Partnering

The GSSD knowledge network is defined as an organized system of discrete actors endowed with knowledge-producing capacity combined through common organizing principles. The actors retain their autonomy; their interaction increases the value of the network to the actors; and the entire knowledge network expands the overall stock of knowledge. Access to interactive knowledge networking enables stakeholder communities to express their preferences and make explicit inputs into decisions. Knowledge metanetworking thus generates new possibilities for empowering individuals and many new modes of interaction among multiple voices ("cyberpartners") that have previously not been heard.

Strategic partnering makes dual outcomes possible: (1) *globalization* of knowledge via greater diffusion; and (2) *localization* of knowledge via representation of local technical and linguistic features. In practice, cyberpartnering can be undertaken in several ways: (1) via mirror siting the GSSD system; (2) distributed knowledge inputs and content provision; and (3) translation to enable multilingual functionality. The current cyberpartners are located in China (Ministry of Science and Technology), the Middle East (American University, Beirut, and ArabDev, Cairo), and France (Ecole des Mines, St. Etienne) and are planned for Central and Latin America as well as Japan and Italy. The institutional needs of cyberpartners can be addressed effectively with very modest resources and can thus significantly enhance diversity on the Internet.

TOWARD A PARADIGM SHIFT

Rapid advances in information technology, the rate at which new users are coming on line, and the growing politicization of environmental and sustainability concerns worldwide suggest that a paradigm shift in the diffusion and use of information technology may be taking place. This change is enabled by technological innovations, but it is driven by powerful synergism generated by a cooperative search for an improved knowledge-management strategy. In its simplest form, the shift is from a paradigm of the unilateral posting of information on the Internet to a paradigm of e-networking and e-conferencing; from centralized to decentralized information management; from uncritical acceptance of information to the evaluation of content and the critical appraisal of the implications; and from one-way communication to multidirectional interaction.

The most important element of this shift by far is the potential leveling of the playing field in knowledge access and management worldwide and its implications for users and providers of knowledge and for demands bearing on the design of new information technologies and systems. The role of GSSD in this changing context can be best understood in terms of its three features: (1) content and connectivity; (2) distributed and decentralized capabilities; and (3) linkages across diverse knowledge and policy communities.

Content and Connectivity

The GSSD strategy focuses heavily on relating *content* (e.g., information, knowledge, data) to *connectivity* (i.e., linkages across topics, subjects, issues, etc., on the basis of substantive meanings) in order to enhance overall *capacity* (performance, choice, and decision). At the simplest level, this means, for example, that Internet resources that address both construction activity and the erosion of agricultural land can easily be taken into account, as relevant, when a search is performed for either one or the other. In practice, these linkages might improve modes of inquiry and/or types of decisions. They would also encourage, if not induce, a more critical appraisal of matters of content.

Distributed Capabilities

The conjunction of new technologies and evolving patterns of information management is turning conventional modes of knowledge development and management upside down. The traditional way of managing information is by centrally controlling operations. The new way is through greater decentralization and the distribution of operational control. The challenge is to ensure reliability and relevance.

The GSSD strategy concentrates on two forms of distributed input (or knowledge provision), knowledge nodes and mirror sites. Both are fundamental to overall system performance. Nodes are knowledge entry points that are slice-specific (topic-specific) and are managed by users with appropriate expertise. Mirror sites function as exact clones of GSSD, with regional input capabilities to ensure the decentralization of the provision of knowledge. Inputs at one mirror site are automatically mapped onto and reflected in the GSSD system as a whole (i.e., in all of the mirror sites). Figure 7 shows key functions.

Linkages across Knowledge and Policy Communities

The underlying premise of the GSSD design is that interconnectivity across knowledge communities generates added value—in other words, the whole is likely to be greater than the sum of its parts—and that the implications of additions, revisions, or changes in knowledge can be better understood if barriers to connectivity across knowledge communities are reduced. The GSSD enables interdisciplinary and multidisciplinary practices to be pursued more easily,

FIGURE 7 Basic layout for Global System for Sustainable Development (GSSD). SOURCE: MIT, 2001.

thereby enabling the generation of new forms of knowledge. The combination of slices and rings and the continuous adaptive updating of knowledge represent the significant facets of the system.

CURRENT STATUS AND CYBERPARTNERSHIP

During the early stages of development, GSSD was implemented on a server written in Mac Common Lisp designed for Macintosh machines (Keene, 1989). The server was built at the Massachusetts Institute of Technology's Artificial Intelligence Laboratory in conjunction with the White House initiative on "re-inventing government." Subsequently, GSSD was adapted to a Lotus Domino Server system that could support a broader user base and enable multilingualism. Other operating modalities currently are being explored for scalability purposes. In reality, all system decisions tend to be shaped by the research environment, not by development from operational imperatives. A research environment is, by definition, experimental, exploratory, and frontier oriented. It encourages new uses and users and new modes of operation, while seeking to "routinize" GSSD capabilities at both local and global levels.

At this writing, the first full version of the GSSD system is in place to support the Global Partnership on Cyberspace for Sustainability. Each slice has been populated with a first round of knowledge (data, analysis, policy experiments, initiatives, strategies, etc.) to test system capabilities in substantive and operational terms. To be effective as a distributed knowledge management system, however, GSSD will require regular maintenance to remove obsolete entries (dead links), add new materials (new links), and, as much as possible, monitor the quality and relevance of the knowledge base to sustainability. Because of their enormous scale and scope, these tasks cannot be centralized; they will require an effective decentralization strategy, which is still experimental and has not been fully articulated. So far, we have proceeded by trial and error. The system's wide-area networking capabilities have also not been realized fully, and critical decisions still have to be made about "best" hardware and software options.

The development of the system has revealed pragmatic challenges to routine maintenance and the need for new decisions. GSSD strategies for nodes and mirror sites are forcing attention on vital operating challenges—over and above those directly related to the research environment. The very nature of advances in information technology and the increase in uses and users is forcing changes in conventional modes of communication and interaction worldwide—in both the scientific and policy domains (Benedikt, 1994). Since the old model of knowledge centralization is being superseded by a new model of distributed knowledge-building and networking, key principles of cyberpartnering are taking shape. These include:

- reinforcing the synergism inherent in the operational division of labor in knowledge provision and information management
- encouraging the sustained decentralization of input capabilities
- reinforcing shared assessments and understandings of quality and quality control
- enabling electronic networking among human and e-networks
- supporting two-way top-down as well as bottom-up communication, nationally and internationally, buttressed by lateral networks

Although technological advances have "pulled" users toward this mode of knowledge sharing and management, the global quest for trajectories toward sustainable development has provided a substantial "push" for a decentralized, distributed, and equal access to cyberspace. This new model is the basis of the Global Partnership on Cyberspace for Sustainability.

The conventional maxim that "knowledge is power" has special implications in this context. The Global Partnership on Cyberspace for Sustainability empowers the scientific and policy communities with the most advanced information technologies. It facilitates collaboration among communities and provides mechanisms for addressing the challenges of sustainability in a coherent and integrated way. All of this improves the possibilities of moving along sustainability trajectories and enhances potentials for global collaboration in the process (Choucri, 1995; UNDP, 1994).

ACKNOWLEDGMENTS

GSSD is the product of collaborative research involving contributions from several research centers at MIT, notably the Technology and Development Program, the Department of Political Science, the Center for International Studies, the Artificial Intelligence Laboratory, and the Laboratory for Computer Science. Fumiaki Shiraishi is responsible for the first round of implementation. The pilot system was developed in collaboration with Juliana Kousoum as initial team leader. I am especially grateful to John Mallery and Roger Hurwitz of the Artificial Intelligence Laboratory and to Gerard McHugh of the Department of Political Science and a researcher in the Department of Ocean Engineering. The first phase adaptation to Lotus requirements was done with Steven Millman and a number of MIT students and researchers, in collaboration with the MIT Intelligent Engineering Systems Laboratory. Farnaz Haghesti is system administrator for GSSD at this writing, with the assistance of Julie Parsons. I am grateful to Paula Deardon for content-provision support and assistance in the routinization of the process. Finally, special thanks to Lutz Gunther-Scheidt (Sony Environment Center, Europe) and Kevin Cavanaugh and Shelby Miller of Lotus Development Corporation. (For a complete list of past and present contributors, see GSSD web site.)

REFERENCES

Alker, H.R., Jr., and P.M. Haas. 1993. The rise of global ecopolitics. Pp. 133–171 in Global Accord: Environmental Challenges and International Responses, N. Choucri, ed. Cambridge, Mass.: MIT Press.

Becker, E., and T. Jahn, eds. 1999. Sustainability and the Social Sciences. Paris: UNESCO.

Benedikt, M., ed. 1994. Cyberspace: some proposals. P. 119 in Cyberspace: First Steps. Cambridge, Mass.: MIT Press.

Choucri, N., ed. 1993. Global Accord: Environmental Challenges and International Responses. Cambridge, Mass.: MIT Press.

Choucri, N. 1995. Globalization of eco-efficiency: GSSD on the WWW. Pp. 45–49 in UNEP Industry and Environment. New York: United Nations Environmental Programme.

Choucri, N., 1999. The political logic of sustainability. Pp. 143–161 in Sustainability and the Social Sciences, E. Becker and T. Jahn, eds. Paris: UNESCO.

Choucri, N., and R.C. North. 1975. Nations in Conflict: National Growth and International Violence. San Francisco: Freeman Co.

Choucri, N., and R.C. North. 1989. Lateral pressure in international relations: concept and theory. Pp. 289–326 in Handbook of War Studies, M.I. Midlarsky, ed. Ann Arbor, Mich.: University of Michigan Press.

Choucri, N., and R.C. North. 1993. Growth, development, and environmental sustainability: profile and paradox. Pp. 67–132 in Global Change: Environmental Challenges and International Responses, N. Choucri, ed. Cambridge, Mass.: MIT Press.

Global System for Sustainable Development. 2001. Massachusetts Institute of Technology. On-line. *http://gssd.mit.edu/.*

Keene, S.E. 1989. Object-Oriented Programming in Common Lisp: A Programmer's Guide to CLOS. Reading, Mass.: Addison-Wesley.

Lang, W., ed. 1994. Sustainable Development and International Law. Boston: Graham & Trotman/ Martinus Nijhoff.

MIT (Massachusetts Institute of Technology). 2001. The Global System for Sustainable Development at MIT. Available online at http://gssd.mit.edu/GSSD/gssden.nsf

Rothenberg, J. 1993. Economic perspectives on time comparisons: alternative approaches to time comparisons. Pp. 355–398 in Global Accord: Environmental Challenges and International Responses, N. Choucri, ed. Cambridge, Mass.: MIT Press.

Tolba, M.K., O.A. El-Kholy, and E.E. Hinnawi, eds. 1992. The World Environment 1972–1992: Two Decades of Challenge. 1st ed. United Nations Environment Programme. New York: Chapman & Hall.

UNDP (United Nations Development Programme). 1994. Technology and Finance: New Opportunities and Innovative Strategies for Sustainable Development. Prepared for the Commission on Sustainable Development Intersessional Working Group on Technology Transfer. New York: UNDP.

Biographical Data

BRADEN R. ALLENBY (cochair) is vice president for environment, health, and safety at AT&T and an adjunct professor at the Columbia University School of International and Public Affairs. Dr. Allenby, a member of the Virginia Bar Association, has represented the Civil Aeronautics Board and the Federal Communications Commission and has been a strategic consultant on economic issues and technical telecommunications. In 1992, he was the J. Herbert Hollomon Fellow at the National Academy of Engineering in Washington, D.C. He was appointed vice president for technology and environment at AT&T's Engineering Research Center in 1994 and subsequently spent two years at Lawrence Livermore National Laboratory as director for energy and environmental systems. Renowned for his pioneering work on industrial ecology, Dr. Allenby is the author of several textbooks on the subject. He has led seminars and workshops and has taught many courses at Yale University School of Forestry and Environmental Studies, the University of Wisconsin Engineering Extension School, and Columbia University. He has also lectured at many other colleges and universities, including Dartmouth, Harvard, MIT, Princeton, Rutgers, the University of California at Berkeley, Stanford, and Tufts. Dr. Allenby is a fellow of the Royal Society for Arts, Manufactures & Commerce. He graduated cum laude from Yale University in 1972, earned a J.D. in 1978 and an M.A. in economics in 1979 from the University of Virginia, and earned an M.A. in 1989 and a Ph.D. in 1992, both in environmental sciences, from Rutgers, The State University of New Jersey.

W. DALE COMPTON (cochair), Lillian M. Gilbreth Distinguished Professor of Industrial Engineering at Purdue University, is a member of the National Academy of Engineering (NAE) and was the first senior fellow of the NAE. After a stint as a professor of physics and director of the Coordinated Science Laboratory at the University of Illinois, Dr. Compton joined the Ford Motor Company, where he was vice president of research. He holds a B.A. from Wabash College, an M.S. from the University of Oklahoma, and a Ph.D. from the University of Illinois, all in physics.

JOHN CARBERRY, director of environmental technology at the DuPont Company, holds a B.S. and M.S. in chemical engineering from Cornell University and an M.B.A. from the University of Delaware. Since joining DuPont in 1965, he has been involved in the development of many chemical processes and new products. In his present assignment, he analyzes environmental issues and recommends technology-based programs and heads a team of scientists working on affordable, publicly acceptable remediation, treatment, and abatement technologies. Mr. Carberry is chair of the National Research Council Panel on Technology for the Disposal of Non-Stockpile Chemical Weapons Materiel, chair of the Chemical Engineering Advisory Board at Cornell University, a fellow of the American Institute of Chemical Engineers, and a registered professional engineer.

NAZLI CHOUCRI is professor of political science and associate director of the Technology and Development Program at MIT, head of the Middle East Program at MIT, and general editor of the *International Political Science Review.* Since joining the faculty of MIT in 1969, her research has focused on sources of conflict in socioeconomic development at the national, regional, and global levels. She is the author of several books, including *Population Dynamics and International Violence* (Lexington Books, 1974) and a companion volume, *Multidisciplinary Perspectives on Population and Conflict* (Syracuse University Press, 1984), that focus on the relationship between population variables and conflict behavior. Dr. Choucri was involved in the preparatory work for the United Nations Conference on Environment and Development in 1992 and participated in the follow-up process leading to the 1997 U.N. General Assembly Special Session, Earth Summit +5. She directed a study on environmental challenges to global accord, the basis for an MIT Press series of which she is the editor and the author of the first volume, *Global Accords: Environmental Challenges and International Responses* (MIT Press, 1995).

Dr. Choucri is currently senior advisor to the heads of two international institutions and has been a consultant to the governments of Abu Dhabi, Algeria, Canada, Colombia, Egypt, France, Greece, Honduras, Kuwait, Mexico, Norway, Pakistan, Qatar, Sudan, Syria, Tunisia, and Yemen. Her current research includes the development and application of advanced information technologies to problems of socioeconomic change and sustainable development.

JULIE E. COHEN, an associate professor of law at the Georgetown University Law Center, teaches and writes about intellectual property law, with a particular focus on computer software and digital works and the intersection of copyright, privacy, and the First Amendment in cyberspace. She is a member of the Advisory Board of the Electronic Privacy Information Center, the Panel of Academic Advisors to the American Committee for Interoperable Systems, and the Committee of Concerned Intellectual Property Educators, a member organization of the Digital Future Coalition. From 1995 to 1999, Dr. Cohen was assistant professor at the University of Pittsburgh School of Law, and from 1992 to 1995 she was an associate with the San Francisco firm of McCutchen, Doyle, Brown & Enersen, where she specialized in intellectual property litigation. She received her A.B. from Harvard-Radcliffe and her J.D. from Harvard Law School, where she was a supervising editor of the *Harvard Law Review*. She is a former law clerk to the Honorable Stephen Reinhardt of the United States Court of Appeals for the Ninth Circuit.

PATRICK D. EAGAN, a program director with the Engineering Professional Development Department of the University of Wisconsin-Madison, has more than 18 years of experience in industry as a design engineer, plant/project manager, business development manager, educator, and researcher. His research interests are focused on design for the environment, environmental supply chain management, and environmentally conscious manufacturing. In addition to developing design tools and courses on industrial environmental design principles and approaches for designers and engineers, he has promoted the accessibility of industrial environmental educational materials to technical professionals here and abroad. Dr. Eagan has lectured and been a consultant on environmental engineering design and "green management" for many companies and institutions, including Johnson & Johnson, Boeing, Eastman Kodak, AMP, the Minnesota Environmental Initiative, Patagonia, and Motorola. He has also served on many panels and committees for the Office of Technology Assessment, the Environmental Protection Agency, the Wisconsin Department of Natural Resources, the University of Wisconsin-Madison, and the Great Lakes/Mid-Atlantic Hazardous Substances Research Center. On the international level, he has been a consultant for Initiativa, GEMI, Austrian Department of Commerce, Korean Foreign Trade Association, Korean Advanced Institute of Science and Technology, and the Pohang University of Science and Technology. Dr. Eagan earned a B.A. in biology from Lawrence University, an M.S. in water resources management from the University of Wisconsin-Madison, an M.S. in civil engineering from the University of Washington, and a Ph.D. in land resources from the University of Wisconsin-Madison.

JERRY FOWLER, a senior scientist at Telcordia Technologies Applied Research, was a member of the technical staff of the InfoSleuth Project at Microelectronics and Computer Technology Corporation (MCC), where he was responsible for the application of InfoSleuth agent technology for NIST's Healthcare Information Infrastructure Technology/Healthcare Enterprise Information

Management Project. Prior to joining MCC, Dr. Fowler was on the faculty of Baylor College of Medicine, where he was involved in research on information management and retrieval and participated in the development of the Virtual Notebook System and the MEDLINE Retriever. He has worked on distributed clinical systems, an immunization registry, and a community health information system and is the author of numerous publications on medical informatics and computer science. He has a B.S. in mathematics and a B.S. in music (string bass) from the University of Oklahoma and an M.S. and Ph.D. in computer science from Rice University.

THOMAS E. GRAEDEL is professor of industrial ecology in the School of Forestry and Environmental Studies, Yale University, a position he assumed after 27 years at AT&T Bell Laboratories. He was the first atmospheric chemist to study the atmospheric reactions of sulfur and the concentration trends for methane and carbon monoxide. As a corrosion scientist, he devised the first computer model to simulate the atmospheric corrosion of metals and was a volunteer consultant to the Statue of Liberty Restoration Project in 1984–1986. One of the founders of the newly emerging field of industrial ecology, he coauthored the first textbook in the field, *Industrial Ecology* (Prentice-Hall, 1995). Mr. Graedel has published nine other books and more than 200 scientific papers.

JOSEPH A. HEIM, a materials engineer, is the head of long-range technology planning at Genie Industries, Redmond, Washington. From 1993 to 1998, he was on the industrial engineering faculty at the University of Washington in Seattle. Before joining the University of Washington, he was senior program officer in the Manufacturing Studies Board of the National Research Council; and from 1990 to 1992, he was the first J. Herbert Hollomon Fellow of the National Academy of Engineering. For several years, Heim was an executive and cofounder of two software product development firms. He has a B.S. in mechanical engineering and an M.S. in computer science from the University of Louisville and an M.S. and Ph.D. in industrial engineering from Purdue University.

JAMES W. HEPTINSTALL was team leader for the redesign of health, safety, and environmental (HSE) processes at Rhône-Poulenc, Inc., at the time of the workshop. His previous corporate-level assignments include a member of the HSE Business Process Redesign Team and team leader, North American Manufacturing Strategy Team (which defined Rhône-Poulenc's long-range manufacturing strategy). He also held management positions at manufacturing facilities of Rhône-Poulenc, Griffin Corporation, and Stauffer Chemical Company. Mr. Heptinstall graduated from Auburn University with a B.S. in chemical engineering.

KOSUKE ISHII earned his B.S.M.E. in 1979 from Sophia University, Tokyo, an M.S.M.E. from Stanford University, and an M.C.E. from the Tokyo Institute of

Technology. After three years as a design engineer at Toshiba Corporation, he returned to Stanford and completed his Ph.D. in mechanical design. He was on the faculty at Ohio State University (OSU) from 1988 to 1994 and is currently an associate professor at Stanford University, where he is codirector of the Manufacturing Modeling Laboratory. His research is focused on life-cycle engineering and robust design. He directs the graduate course sequence on designing for manufacturability on the Stanford Instructional Television Network. Dr. Ishii is the author or coauthor of more than 80 refereed articles, was chair of the ASME Computers in Engineering Division in 1998, and was an associate editor of the *Journal of Mechanical Design* and *AI in Engineering*. He is the recipient of many awards, including the Lilly Fellowship for Excellence in Teaching (1989), the National Science Foundation Presidential Young Investigator Award (1991), OSU Lumely Research Award (1992), Pitney Bowes-ASME Award for Excellence in Mechanical Design (1993), OSU Harrison Faculty Award (1994), AT&T Industrial Ecology Faculty Fellowship (1995), General Motors Outstanding Long Distance Learning Faculty Award (1996), and the LG Electronics Advisory Professorship (1997).

MICHAEL R. KABJIAN has been a consultant to a wide range of clients around the world on environmental and chemical management information systems. Prior to striking out on his own, he was a project manager for EMAX Solution Partners, where he was responsible for deploying information management systems for environmental compliance in various manufacturing facilities. He has also designed and developed information systems for Roy F. Weston, Inc., to support corporate initiatives in designing for the environment, life-cycle analysis, and risk assessment. Mr. Kabjian holds a B.S. in environmental systems engineering and a B.S. in strategic management from the Wharton School of Business, University of Pennsylvania.

PAUL C. KILLGOAR, Jr., has been manager of vehicle crash safety research, Product Development Center, Ford Motor Company, since 2000. His areas of responsibility include advanced work on vehicle safety systems and the verification of safety performance of new vehicles. He received his B.S. in chemistry from Bridgewater State College and a Ph.D. in physical chemistry from Michigan State University in 1972, when he joined the Polymer Science Department of Ford Research. In 1991, he became manager of research programs on fuels and lubricants for power-train applications. The major foci of this research were the liquid-phase oxidation of hydrocarbons, tribology, and the volatility behavior of fuels. From 1992 to 1995, Dr. Killgoar was manager of the chemistry research department, where he directed major projects on the gas-phase photochemistry of air pollutants, the modeling of atmospheric pollution processes, the real-time measurement of emissions from vehicles, and the control of manufacturing emissions, including bioremediation. From 1995–2000 he was manager of the Manufacturing Systems Department of Ford Research. The research activities of the department cover most manufacturing disciplines, such as painting, machining, joining, rapid prototyping and tooling, plastics processing, and sheet-metal forming. He has been granted eight patents and has published 24 papers on coating, adhesives, elastomers, polymer processing, and related topics.

PAUL R. KLEINDORFER is the Universal Furniture Professor of Decision Sciences and Economics and professor of public policy and management at the Wharton School of Business, University of Pennsylvania. He is codirector of the Wharton Center of Risk Management and Decision Processes, where he conducts research on regulated industries, focused on energy and environmental problems. Dr. Kleindorfer also recently initiated the Infrastructure Forum, in cooperation with the Interdisciplinary Center for the Study of Business, Law, and Technology, in Herzliya, Israel, to promote research and education on infrastructure development and commercialization in the Mediterranean region. He has held university appointments at Carnegie Mellon University (1986–1989), Massachusetts Institute of Technology (1969–1972), Wharton School of Business (1973–present), and several international research institutes, including IIASA (Vienna) and the Science Center (Berlin). He is the author or coauthor of 10 books and more than 100 research papers on managerial economics, productivity, and regulation. He has been a consultant to the U.S. Postal Service and many regulated companies in the telecommunications and energy sectors. Dr. Kleindorfer graduated with distinction with a B.Sc. from the U.S. Naval Academy in 1961. He studied on a Fulbright Fellowship in mathematics at the University of Tubingen, Germany (1964–1965) and pursued doctoral studies at Carnegie Mellon University, where he received his Ph.D. in 1970 in systems and communications sciences.

JOSHUA KNAUER is one of the leading (and youngest) pioneers of progressive on-line activism. As a student at Carnegie Mellon University in 1991, Mr. Knauer started the EnviroLink Network, now the Internet's largest environmental information resource for activists, organizations, businesses, and government. Mr. Knauer created GreenMarketplace.com in 1999, an e-commerce site for socially and environmentally responsible products, services, and information. GreenMarketplace.com and EnviroLink have generated a great deal of media attention, including a prominent write-up in a *Time Magazine* special issue, "Heroes for the Planet." Mr. Knauer has also appeared as a commentator on socially and environmentally responsible business practices for the Fox News Channel. He was recently named one of AlterNet's "New Media Heroes," and his writings have appeared in numerous books and journals.

Mr. Knauer is a member of the boards of directors of several organizations, including Sea Shepherd Conservation International, Institute for Global Communications, Allegheny Sierra Club, and EnviroLink. He frequently makes presentations at conferences and speaks to high school and college students about balancing social, environmental, and economic values. Mr. Knauer received a B.S. in environmental ethics and policy from Carnegie Mellon University.

DAVID S. LIEBL is an associate faculty member of the Department of Engineering Professional Development Program at the University of Wisconsin-Madison and a waste reduction and management specialist with the Solid and Hazardous Waste Education Center at the University of Wisconsin-Extension. In 1996, he chaired the National Pollution Prevention Roundtable's Information and Technology Transfer Working Group, which led the effort to establish a national pollution prevention information network (*http://www.p2.org*). In collaboration with pollution prevention programs in the Great Lakes region of the United States and Canada, he established the Great Lakes Technical Resource Library. He also founded the P2Tech pollution prevention listserv (*http://www.great-lakes.net/lists/p2tech*) and created the VENDINFO directory of pollution prevention technology manufacturers and service providers (*http://es.epa.gov/vendors*).

WILLIAM M. MARTIN is the senior law clerk to the Honorable Joseph F. Weis, Jr., of the United States Court of Appeals for the Third Circuit. At the time the article was written, Martin was a student at the University of Pittsburgh School of Law and a managing editor of the *University of Pittsburgh Law Review*. Martin will be joining the Antitrust Division of the U.S. Department of Justice in the fall of 2001.

GREG PITTS is the executive director of Ecolibrium, a nonprofit research and education organization. At the time of this writing, Mr. Pitts was the director of environmental programs at Microelectronics and Computer Technology Corporation, where he was responsible for environmental technology development and for addressing the effect of environmental issues on competitiveness in the electronics and computer industries. He established a program to direct several research projects on the development of environmentally friendly methods of fabricating printed wiring boards and the disposition (e.g., reuse, remanufacture, recycling, etc.) of electronic products, designing environment tools, and promoting electronic access and use of environmental information. He managed an industry/government/academia life-cycle assessment of a computer work station and a study of environmental issues for the electronics industry.

DEANNA J. RICHARDS is an independent consultant on environmental and sustainable development. From 1990 to 1999, Dr. Richards directed the National Academy of Engineering (NAE) Program on Technology and Environment/Technology and Sustainable Development. From 1996 to 1998, she served as acting director of the NAE Program Office and oversaw the management of six additional programs. During her tenure at NAE, she launched the institution's initiatives on industrial ecology and engineering ecological concerns. Before joining the NAE, Dr. Richards was an assistant professor of environmental engineering at the University of North Carolina at Charlotte. Her research and publications were focused on advanced biological treatment systems for the treatment of hazardous waste. A registered professional engineer, she also worked for several years as an environmental engineer in the United States and Malaysia. She received a B.S. (honors) in civil engineering from the University of Edinburgh in Scotland and an M.S. and Ph.D. also in civil/environmental engineering from the University of Pennsylvania.

MAURICE RICKARD was the creative director of the EnviroLink Network, the new-media editor of the *electronic book review*, and a media consultant at the time of the workshop. He has worked with Internet and print-based electronic publishing since 1986, designing, writing, proofreading, and producing web sites, catalogs, promotional brochures, newsletters, and other publications and developing and implementing electronic publishing systems. He has been a consultant to Graphical Arts Technical Foundation, Carnegie Mellon Research Institute, Black Box Corporation, Westinghouse Science and Technology Center, HealthAmerica, the University of Pittsburgh Press, the Art Institute of Pittsburgh, and other organizations. Rickard has an M.A. in English from the University of Pittsburgh.

ELI M. SNIR is a lecturer in the Operations and Information Department of the Wharton School of Business, University of Pennsylvania, where he recently completed his Ph.D. As a research fellow at the Wharton Risk Management and Decision Processes Center, Dr. Snir's research focused on evaluating the importance of information sharing in supply chains as well as the effect of liability-sharing rules on environmental policies and procedures. His recent research has been focused on the outsourcing of information technology services. Before his enrollment at Wharton, Dr. Snir was an officer in the Israeli Army for five years. He received an M.Sc. in operations research and systems analysis and a B.Sc. in industrial engineering from the Technion-Israel Institute of Technology.

LYNDA M. WIESE was director of the Bureau of Cooperative Environmental Assistance with the Wisconsin Department of Natural Resources (WDNR), which examines nontraditional approaches to waste reduction and pollution prevention through partnerships with industry and innovative approaches to pollution control, such as ISO 14000 and whole-facility regulation. As a field engineer and supervisor for the WDNR Air Management Program, she spent 14 years working on regulatory compliance, regulatory enforcement, and permitting. Ms. Wiese was involved in policy development through the Wisconsin Clean Air Act Task Force and was chair of the WDNR Asphalt Paving Technology Transfer Team. She graduated from Northland College in Ashland, Wisconsin, with a B.S. in geology and meteorology. Ms. Wiese, who died of cancer in September 2000, is sorely missed by her family, friends, and colleagues.

Index